How do we decide whether a particular vertebrate species can be classified as a pest? What is achieved by control of such pests? The approach advocated by Jim Hone uses statistical and economic analyses as well as mathematical modelling to determine the pest status of species and to analyse the effects of control on pest abundance and on the damage caused by pests. To do this the book reviews critically the literature on damage assessment and control evaluation. Links are then made to relevant topics in ecology, epidemiology, fisheries management and economics, showing how analyses in other scientific fields can be used in the analysis of vertebrate pest control.

The scope of the book is worldwide and many mammalian and avian pests are described using case studies. The emphasis is on the evaluation of data, rather than on specific control methods.

Analysis of vertebrate pest control

Cambridge Studies in Applied Ecology and Resource Management

The rationale underlying much recent ecological research has been the necessity to understand the dynamics of species and ecosystems in order to predict and minimise the possible consequences of human activities. As the social and economic pressures for development rise, such studies become increasingly relevant, and ecological considerations have come to play a more important role in the management of natural resources. The objective of this series is to demonstrate how ecological research should be applied in the formation of rational management programmes for natural resources, particularly where social, economic or conservation issues are involved. The subject matter will range from single species where conservation or commercial considerations are important to whole ecosystems where massive perturbations like hydro-electric schemes or changes in land use are proposed. The prime criterion for inclusion will be the relevance of the ecological research to elucidate specific, clearly defined management problems, particularly where development programmes generate problems of incompatibility between conservation and commercial interests.

Also in the series

Graeme Caughley, Neil Shephard & Jeff Short (eds.) *Kangaroos: their ecology and management in the sheep rangelands of Australia*
P. Howell, M. Lock & S. Cobb (eds.) *The Jonglei canal: impact and opportunity*
Robert J. Hudson, K.R. Drew & L.M. Baskin (eds.) *Wildlife production systems: economic utilization of wild ungulates*
M.S. Boyce *The Jackson elk herd: intensive wildlife management in North America*
Mark R. Stanley Price *Animal re-introductions: the Arabian Oryx in Oman*
R. Sukumar *The Asian Elephant: ecology and management*
K. Homewood & W.A. Rodgers *Maasailand ecology: pastoralist development and wildlife conservation at Ngorongoro, Tanzania*
D. Pimentel (ed.) *World soil erosion and conservation*
R.J. Scholes & B.H. Walker *An African savanna: synthesis of the Nylsvley study*
T. D. Smith *Scaling fisheries: the science of measuing the effects of fishing 1855–1955*

ANALYSIS OF VERTEBRATE PEST CONTROL

Jim Hone
Applied Ecology Research Group, University of Canberra, Australia

CAMBRIDGE
UNIVERSITY PRESS

Published by the Press Syndicate of the University of Cambridge
The Pitt Building, Trumpington Street, Cambridge CB2 1RP
40 West 20th Street, New York, NY 10011-4211, USA
10 Stamford Road, Oakleigh, Melbourne 3166, Australia

First published 1994

Printed in Great Britain at the University Press, Cambridge

A catalogue record for this book is available from the British Library

Library of Congress cataloguing in publication data

Hone, Jim.
Analysis of vertebrate pest control / Jim Hone.
p. cm. – (Cambridge studies in applied ecology and resource management)
Includes bibliographical references (p. 220) and index.
ISBN 0 521 41528 4 (hardback)
1. Vertebrate pests – Control – Evaluation – Statistical methods. 2. Vertebrate pests – Control – Mathematical models. 3. Vertebrate pests – Research – Statistical methods. 4. Vertebrate pests – Mathematical models. I. Title. II. Series.
SB993.4.H66 1994
632'.66–dc20 93-49727 CIP
ISBN 0 521 41528 4 hardback

VN

CONTENTS

PREFACE

The idea for this book developed out of my interest into how other scientists had estimated the damage by vertebrate pests and the effects of vertebrate pest control. How did scientists obtain data and how did they analyse it? Was it of any use? In writing this book I have attempted to collate and critically comment on information on a variety of topics in vertebrate pest control. The collation is not exhaustive. Many good studies have not been used simply because of limited space.

I thank many people for assistance with discussions and comments on draft manuscripts. Graeme Caughley and Tony Sinclair of the editorial board of the books in this series assisted with discussions and negotiations. Roger Pech, Peter O'Brien, Glen Saunders, David Choquenot, John Parkes, Graham Nugent, Chris Frampton, Clem Tisdell, Peter Whitehead, Mary Bomford, Mike Braysher, George Wilson, Peter Brown, Chris Cheeseman, Stephen Harris and Astrida Upitis provided useful comments. The ideas and details are, however, mine, so blame me for any errors or omissions. Fellow staff and students at the University of Canberra tolerated my occasional absences to write the book. Staff at the Central Science Laboratory, Worplesdon, Surrey, provided discussions and facilities. I am grateful to them all. I also thank Alan Crowden and Alison Litherland of CUP for editorial assistance.

Finally, I am an ecologist, not a biometrician, economist or mathematician. I have attempted to bring their fields of study a bit closer to my own and that of my fellow ecologists. I hope that fellow ecologists can, and do, use the results, though they should not expect to be spoon-fed. In the words of a much greater author, 'I should not like my writing to spare other people the trouble of thinking. But, if possible, to stimulate someone to thoughts of his own' (Wittgenstein, 1967).

1

Introduction

Pests can be defined as organisms that cause harm: economic, environmental or epidemiological. Cherrett *et al.* (1971) defined a pest animal as one which is noxious, destructive or troublesome to humans. Woods (1974) qualified the definition to incorporate a requirement of causing economic damage. Today, vertebrates cause problems in agriculture: to crops, to livestock, in and around buildings and other equipment and by spreading diseases. Problems also occur in forestry, in conservation of plant and animal species and communities, in urban industry and in our own homes.

There are difficulties, however, in defining a pest (Harris, 1989). Is it that coyotes kill sheep or the number of sheep killed that causes coyotes to be called a pest? Answering this question is partly a social issue but also a scientific issue. This book is about the science of evaluating vertebrate pest control with particular emphasis on mammals and birds as pests; how to analyse whether populations of vertebrates really are pests and how to plan and assess what is achieved by pest control. It is not, however, about how to control vertebrate pests. For example, the book is not about how to control rats and mice, but describes how to determine if they are pests, and if so, how to estimate what is achieved by control.

Worldwide in scope, the book describes and critically reviews the literature on a range of analyses used in vertebrate pest research and management. The review draws on theories and analyses used in other scientific disciplines, shows their relevance to vertebrate pest control and how they are or can be utilised. In particular, predator–prey theory in ecology, and the theory of disease spread in epidemiology, are examined.

Statistical, economic and modelling analyses are described, with examples, that cover the spectrum of damage by vertebrate pests, and the methods used to control these pests. The analyses help identify the effects of pests, examine the ecological and economic levels of those effects and analyse the effects of control on the pest populations and their impacts. Hence, the book will assist in pest control decisions that have a scientific or economic basis. The details of the use of control methods, such as poisoning, trapping, shooting, fencing and habitat manipulation are covered in many references, for example, Timm (1983) and Breckwoldt (1983). The control methods used can be influenced by practical experience, as well as broader social, legal and technological factors. For example, the enthusiasm of some people for pest control may be related to some dark overtones in history reflected in the origin of the word 'pest' from the Latin word '*pestis*' meaning plague or contagious disease (Cherrett *et al.*, 1971).

A distinction between the effects of control on pests and on damage is fundamental to the book. I have observed that societies are more interested in the economics of damage control than the statistics of pest control. My reading of the literature is, however, that scientists have the reverse priorities. One implication for writing this book is that there is vastly more information available on the statistics of pest control than the economics of control, so the chapter 'Economic analysis' is thinner than I would have liked. There are simply fewer examples to write about.

The book is structured to answer several questions. Firstly, does the species of concern actually cause any damage (Chapter 2)? If it does, how can the effects of subsequent control be estimated, in terms of the level of impacts and abundance of the pest? This is described in Chapter 3. Is control economical? This question is addressed in Chapter 4. How can we model the population dynamics of pests and their damage? These are described in Chapter 5, while Chapter 6 describes how to model control, the process itself and its effects on pests and damage.

The conceptual framework used (Table 1.1) describes the variables estimated by research and the types of analysis. The variables estimated are usually one, or more, of the level of pest damage, pest abundance, pest population rate of increase or some other demographic parameter, such as sex ratio or age structure. There are three broad classes of analysis: statistics, economics and modelling.

The structure of the book is not by species or even by groups of species, for example rats, birds, ungulates, but by analyses and management topics. I could have structured the book according to pest species but that would have overemphasised the species rather than the analyses. Hence, the book is entitled *Analysis of Vertebrate Pest Control* and not *Vertebrate Pest Control*. The effect of the structure is to allow the reader to explore the use of a particular analysis on their species of interest and easily locate nearby a review of other uses of that analysis on other pest species. For example, information on statistical analysis of rat damage is close to a review of statistical analysis of bird strikes on aircraft. To aid readers, discussion of related topics is cross-referenced, both forwards and backwards, within the text.

There are six topics that are discussed in each of Chapters 2 to 6. The topics are predation of livestock, infectious diseases, rodent damage, bird strikes on aircraft, bird damage to crops and rabbit damage. In some chapters additional topics are discussed, such as predation of rock-wallabies by foxes in Chapter 3 and non-target effects of control in Chapters 3 and 6. The generic aspects of several particular methods of vertebrate pest control, such as poisoning, biological control and shooting, are discussed in Chapters 3 and 6.

Experimental design and sampling are emphasised in the discussion of statistical analysis. However, this is not a book on statistics. A knowledge of basic statistics and mathematics is assumed, but references are provided where more detailed analyses are used. I have tried to avoid using statistics solely as a line-fitting exercise without regard to the biology and practical application of what is being studied. Macfadyen (1975) expressed a similar concern in ecology. Emphasis is on the relevance and outcomes of statistical, economic and modelling analyses, rather than on the analyses themselves. Hence, detailed examples of analysis of variance or cost–benefit analysis are generally not provided. More details of the steps in and predictions of mathematical models are given, however, as fewer readers may be familiar with such details. The chapters on modelling are primarily about using mathematics to test hypotheses, not about mathematics itself. Hypothesis testing is needed in vertebrate pest control as it is needed in ecology (Macfadyen, 1975).

Table 1.1. *Variables estimated in studying vertebrate pest control and the analyses used to estimate damage or effects of control. The body of the table lists relevant chapters in the book*

	Analyses		
Variables estimated	Statistics	Economics	Modelling
Damage	Chapter 2	Chapter 4	Chapter 5
Response to control			
(i) Damage	Chapter 3	Chapter 4	Chapter 6
(ii) Abundance, rate of increase and other demographic parameters	Chapter 3	Chapter 4	Chapter 6

The book is a critical review of the literature. Hence, for the sake of balance I have included discussion of the strengths and weaknesses of studies. For example, some studies have used sophisticated experimental designs and others have used no statistical analysis. I have generally interpreted the lack of statistical analysis, particularly in manipulative experiments, as a study weakness, for reason that will be apparent within the book.

Pest damage and the effects of control can be studied by empirical and theoretical methods. The empirical methods are described first (Chapters 2, 3 and 4), then the theoretical methods, particularly mathematical models, are described (Chapters 5 and 6). The final chapter is a brief summary and outlook. The book is specifically written for research workers, academics, professional wildlife managers and undergraduate and postgraduate students with a focus on wildlife management. To accommodate this broad readership, the book is divided into chapters with differing levels of assumed prior knowledge. Chapters 2, 3 and 4 assume a minimum of mathematical knowledge. Chapters 5 and 6 are more mathematical and may be less accessible to some. I have tried to illustrate the ideas both graphically and in equations in Chapters 5 and 6. Some readers may be daunted by an equation yet easily grasp the idea when shown a diagram.

The analyses concentrate on populations of vertebrate pests rather than individuals. For example, the trapping of rats or foxes focusses the attention of the trapper on the individuals. Here, we are more concerned with the effects of trapping on a population of rats or foxes.

Research on vertebrate pests can be difficult because the animals range over large areas and the research is affected by the weather and other factors not under the direct control of the researcher. Some of those problems also occur in agricultural research (Dillon, 1977) so the similarities are examined. In particular, difficulties relate to the use of experimental design, statistical estimation, response variability, economics of research, differences between experimental and field results and making recommendations to farmers and wildlife managers. These topics are discussed in many sections of the book with suggested solutions.

The 'conventional wisdom' of vertebrate pest control is that if animals cause damage, then reducing animal abundance by control, will reduce damage. Deeper analysis is needed to examine this logic as this book is not about accepting 'conventional wisdom'. For example, is the relationship between animal abundance and the level of damage linear or non-linear? If the relationship is linear then a reduction in pest abundance will reduce their impacts. If the relationship is non-linear a lot of control may not produce a substantial change in damage, but slightly more control will significantly reduce damage. The answer is embedded in the relationships between the level of control and the reduction in pest abundance and the level of control and the reduction in pest damage. The relationships may be easily extricated or may require very detailed study. Estimation of the relationships will allow scientists to answer questions such as: 'what level of control will reduce pest damage by 70%, or what level of control is needed such that the economic benefits of control exceed the economic costs of control?' Dyer & Ward (1977) described the need for more critical analysis and interpretation of control of bird pests, especially the relationship between bird abundance and the level of damage. Similarly, the level of control that maximises net economic benefits may, or may not, be obvious. These issues are addressed in this book.

Planning of vertebrate pest control is no different from planning other human activities. It involves deciding on objectives, alternative strategies, implementing one or more of those strategies and evaluating the results. If the results achieve the objectives then no further action may be necessary, but if the objectives are not attained then the actions or the objectives may need to be changed. A useful planning framework that is also suitable for vertebrate pest control was described by McAllister (1980). The framework, as applied to pest control, emphasises the need to evaluate alternatives and monitor the results. Each step needs some criterion as a basis for decision-making. Are alternatives selected on the basis of the highest pest mortality, lowest cost or highest benefits per unit cost? The mortality statistics are discussed in more detail in Chapter 3 and the economic criteria in Chapter 4. The strategies may be designed and evaluated by experience, laboratory and field testing, and by modelling. The latter is discussed in Chapters 5 and 6. The selection of alternatives can use pre-selected criteria. Dyer & Ward (1977) described flow diagrams for such decisions in the control of bird pests and their impacts.

Bell (1983) described a framework for planning and implementing options for field management of large mammals in African conservation areas. The discussion is useful although not specifically concerned with pest control. In management planning two types of decision were distinguished – technical and preferential. Technical decisions are those dealing with facts, such as how to design a trap to hold a certain size animal. Preferential decisions are those not based on fact but rather on opinion or preference, such as deciding whether to use lethal or non-lethal control methods. The information for making decisions was recognised as being inadequate. The decision to cull animals because of 'overabundance' implies two decisions – a preferential decision that acceptable limits to grazing or damage exist, and the technical decision that these limits have been exceeded. The analyses described in this book will provide inputs to such technical decisions.

Caughley (1981) identified four classes of 'overpopulation' of mammals. The first class was of animals that threaten life or livelihood. The second class was of animals that depress the density of favoured species. The third class was of animals that were too numerous for their own good, and the fourth class was of animals in systems off their equilibrium. The first two classes are those in which vertebrate pests usually occur. The analyses in this book will help determine whether 'overpopulation' means the animals are pests. For example, introduced reindeer (*Rangifer tarandus*) are abundant on South Georgia island (Leader-Williams, Walton & Prince, 1989) and have changed some of the native vegetation, but not caused any extinctions of species. That suggests they should be classified in the second class of pest. However, Leader-Williams *et al.* (1989) argued there was no scientific basis for active pest management, apparently as no species were threatened with extinction. This example illustrates the difference between technical decisions on how much control is needed, and the preferential decision on whether any control is needed.

A related question is should a pest be controlled or eradicated? Legislation may require eradication but it may not be economically feasible. The strongest ecological or economic arguments for eradication probably occur for pests on small islands or in small continental areas (Pimm, 1987). The decision to eradicate a pest should involve prior analysis and not be taken lightly. Coypu (*Myocastor coypus*) eradication from Britain was preceded by evaluation of a trial eradication in East Anglia (Gosling, Baker & Clarke, 1988), as described in section 3.17, and by simulation modelling of coypu dynamics and of the trapping effort needed (Gosling & Baker, 1987), as described in section 5.3.1. The subsequent eradication programme was based on trapping and trapper incentives (Gosling & Baker, 1989).

According to Kim (1983), eradication programs cannot be successful unless five conditions are met: (i) an appropriate control method exists, (ii) basic information on biology and ecology are available, (iii) defined natural barriers exist which stop pest movement, (iv) cost–benefit analysis of alternative actions has occurred, and (v) organisations are prepared. The conditions were based mostly on evaluation of eradication of exotic plant pests in northern America. The eradication of coypu in England met each condition, except (iv) – there was no cost–benefit analysis – so not all conditions are essential. Eradication of feral goats from New Zealand was not considered feasible because of the likelihood of domestic escapes (Parkes, 1990), so condition (iii) would not be met.

The five points of Kim (1983) can be reduced to one: the rate of increase of the pest population must be negative. A population must go extinct if its rate of increase stays negative. Extinction can be because of eradication (extreme control) or because the animal's habitat has insufficient resources as a result of human action. Methods for estimating rate of increase are described in section 3.4.

The decision to control or eradicate should be based on an assessment of the relationship between damage and pest abundance, as described further in Chapters 2 and 5. If the vertebrate species does not have a direct impact on human activities but influences it indirectly through its role as a host for a disease, then eradication should not automatically be an objective. Such a situation can occur with foxes and rabies. The maintenance of the infection depends on the abundance of the hosts and the contact rates between them (Bailey, 1975; May & Anderson, 1979). This is discussed in more detail in section 5.6.

To sum up, this book is about analyses used to study vertebrate pest control. Emphasis is on statistical, economic and modelling analyses and their application. Emphasis is not on the specific methods of vertebrate pest control or on the pest animals themselves. The first set of analyses, statistical analysis, is now explored. The estimation and statistical analysis of the type and level of damage is the subject of the next chapter.

2

Statistical analysis of damage

Imagine a scene of many square kilometres of semi-arid shrubland where an isolated population of 1000 goats is eating the local shrubs. The scene could be interpreted in many ways. In their native environment, or one long-managed for agriculture, the goats may be called indigenous or domestic animals respectively. In other parts of the world the goats may be called pests. What is the difference? It is the effect that the goats have on the shrubland rather than the goats themselves that can be annoying and may necessitate pest control. Hence, some scientific assessment of the effects of the goats is needed.

In a broader context, it is the damage of vertebrate pests that justifies their economic control. It is often the origin (indigenous or exotic) or legal status (declared pest) of a species that is the social justification for control. Non-native species are often labelled as pests without careful assessment of their pest status; are their effects really damage? The differentiation between the impact of a species and the species itself is fundamental to control and to the topics reviewed in this book. This chapter is concerned with estimating the damage of vertebrate pests. Emphasis is on the statistical aspects of design and sampling, with case studies used to illustrate the principles. Finally, there is a review of the analyses used.

For effective pest control it is fundamental to determine the pest status of an animal. Until this is established, a statistical or economic evaluation of damage is pointless. Judenko (1973) listed eight reasons for assessing the losses caused by pests: (i) establish the economic status of specific pests, (ii) estimate the damage or abundance level that justifies control, (iii) calculate the justifiable expenditure on control, (iv) estimate the effectiveness of control, (v) measure the effects of environmental factors on the loss of yield caused by pests, (vi) give information to manufacturers and distributors of pesticides to help them decide what actions should be taken, (vii) assess the public use of funds for current research, and (viii) direct future research and planning.

Auld & Tisdell (1986) expanded on points (iii) and (vii), by arguing that it is only rational to increase our knowledge about pests up to the point where the marginal benefit from research equals the marginal cost of obtaining that extra knowledge. Methods of estimating such marginal benefits and costs are described in section 4.2. Research into rodent control in Britain during the 1940s was criticised by Barnett & Prakash (1976) when they asked what the final effects of the rodent control were on food saved and abundance of rodents. They considered that few researchers had tried to answer such a question.

2.1 Types of damage

The damage by vertebrate pests is varied and the reported or alleged impacts are probably even more diverse. The impacts include effects on soil structure, soil erosion, water quality and runoff, alteration of plant species composition, changes in plant growth, biomass, reproduction and crop production, changes in animal populations and production, spread of human and livestock diseases, changes in ecosystem structure, and even species extinction. Pimm (1987), Brockie *et al.* (1988) and Ebenhard (1988) review damage caused by introduced pest species around the world.

2.2 Spatial and temporal variation

The damage by vertebrate pests can vary in space and time. Emphasis in vertebrate pest research is, however, often on the mean level of damage, but we need to examine the variation about the mean and the sources of such variation. This section examines the patterns in such variation.

Spatial variation is the frequency distribution of damage between areas at the same time. This variation can be examined at different geographic scales, such as within fields or forests, or in a district or county, or a region or country. The distribution may vary from negative exponential to normal to positively skewed (Fig. 2.1). Each unimodal (one-peaked) frequency distribution is an example of a more general exponential frequency distribution (Dobson, 1983). Bimodal (two-peaked) or multimodal (many-peaked) distributions could also occur (see section 3.1 and Fig. 2.1), though none have been reported as far as I know.

Dawson (1970) reported a positively skewed frequency distribution of damage to grain crops by sparrows (*Passer domesticus*) in parts of New Zealand, and Buckle, Rowe & Husin (1984) cited a study of rat damage to rice fields in Malaysia that showed a log-normal frequency distribution of damage. Negative exponential frequency distributions of damage have been reported for damage by feral pigs (*Sus scrofa*) in south-eastern Australia (Hone, 1988a), damage to sunflowers in North and South Dakota as a result of birds, particularly blackbirds (*Agelaius phoeniceus*, *Xanthocephalus xanthocephalus*) and grackles (*Quiscalus quiscula*) (Hothem, DeHaven & Fairaizl, 1988), and damage to grapefruit by grackles (*Quiscalus mexicanus*) in Texas (Johnson, Guthery & Koerth, 1989).

The pattern of spatial variation in damage is caused by, or correlated with, many factors, including the behaviour of pests or the characteristics of the environment. Harmon, Bratton & White (1983) reported that rooting by wild boar (*Sus scrofa*) in Great Smoky Mountains National Park in Tennessee was most frequent in gray beech forest at 1450 to 1800 m elevation. Rooting of the ground by feral pigs in mountain forests and woodlands in south-eastern Australia was positively correlated with elevation (Hone, 1988b).

Otis & Kilburn (1988) reported nearby marshes, roosting sites for birds, had the largest influence on blackbird damage to sunflowers in North and South Dakota and Minnesota. Glahn & Otis (1986) analysed, by categorical modelling, factors influencing blackbird and starling damage at livestock feeding operations in Tennessee. Blackbird damage was related to feeding of domestic animals, particularly pigs, on the ground, and the presence of a large

Fig. 2.1. Possible frequency distributions of spatial variation in damage by vertebrate pests. (*a*) Negative exponential, (*b*) normal, (*c*) positively skewed, (*d*) bimodal. In each example the *x* axis gives the level or extent of damage, and the *y* axis gives the number of sites with each level of damage (the frequency).

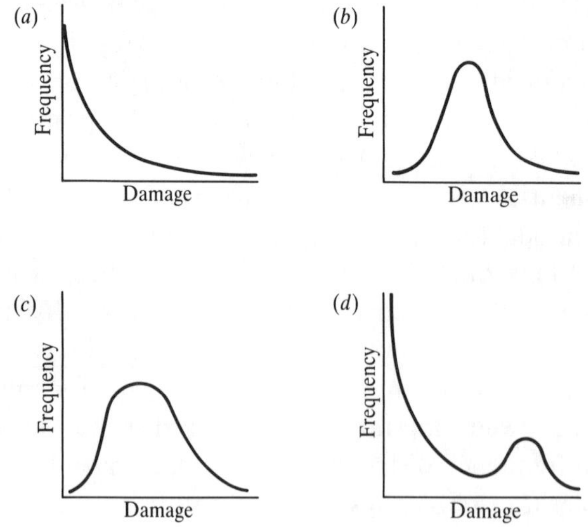

roost within 15 km of the farm. In contrast, weather was the main factor associated with starling damage, though many other environmental and farm features were also statistically significant. The effect of grazing by rabbits (*Oryctolagus cuniculus*) on winter wheat in southern England was studied by Crawley & Weiner (1991). Continuous grazing changed the shape of the frequency distribution of plant size and many other yield components. Changes occurred to the means, variability, skewness and kurtosis (the pointedness of the distribution).

Scale also affects the estimation of damage. For example, if one was estimating the effect of rabbits on pasture biomass in a field then two situations could occur. Firstly, all plots on which biomass was measured could occur within the home range of the rabbits. Secondly, if the plots are selected at random, then some may be outside the rabbits' home range so are never visited. The first study would estimate the effect of rabbits on biomass within the home range of the rabbits, while the second study would estimate the effect on the whole site, which would be expected to give a lower estimate. The second estimate is not biased, but is simply estimating a different effect because of the different scale of the study. If rabbits did decrease pasture biomass then the second study would have a lower probability of detecting the effect. Obviously, experimental control plots which had no rabbits would be needed in both situations.

Temporal variation is the difference, between times, in damage at a site. The frequency distributions of damage could be similarly varied as for spatial variation (Fig. 2.1). Time can influence the impact of vertebrate pests in several ways in a similar manner to the effects of time on agricultural production (Dillon, 1977). Firstly, time may directly influence production and hence act as an input factor. In the same way, the responses of vegetation to exclusion of herbivores may be variable in semi-arid and arid environments because of variable rainfall. Secondly, the level of input variables, such as the number of pests, may vary over time; hence time acts indirectly on damage. Thirdly, the time sequence of inputs may influence damage. For example, 100 pests early in the life of a crop and 10 later near harvest, may have less impact on a crop than 10 early and 100 nearer harvest. Fourthly, there may be carry-over effects on impacts. For example, crop yield may be partly related to effects of fertilisers applied to a crop sown in the previous year. Therefore the effects of time may not be independent. Lastly, a crop may be able to compensate for some continuous damage but after a certain duration such damage is accentuated.

Temporal variation includes the cumulative damage of a pest, for example, damage to a crop. Halse & Trevenen (1985) estimated that the percentage of leaf area of medic (*Medicago* spp.) removed by skylarks (*Alauda arvensis*) in north-western Iraq increased over time. Unfortunately, this damage was not expressed in physical or production units. Gillespie (1982) reported a similar curvilinear increase in damage by greenfinches (*Carduelis chloris*) to a rape crop in southern New Zealand. Cummings, Guarino & Knittle (1989) showed that most blackbird damage to sunflowers in North and South Dakota and Minnesota occurred within 18 days of flowering (anthesis) by the plants. Damage to oilseed rape crops by wood-pigeons (*Columba palumbus*) in central and southern England was highest in March and lowest in December (Inglis, Thearle & Isaacson, 1989). Damage in December or April had no significant effect on yield, as assessed by two-factor analysis of covariance.

Recognition of temporal variation is necessary for planning the duration of a study to estimate pest impacts. If temporal variation is low then a short study may be adequate, but if temporal variation is high then a longer study may be needed. Hairston (1989) recommended that the duration of a study should be determined at the start, not along the way. The latter approach leads to the temptation to stop the study when the results are pleasing to the scientists or others associated with the study.

2.3 Evaluation of damage

The assessment of pest status involves direct observation of the activities of the suspected pest, or if that is not possible then a comparison of the level of impact with and without the suspected pest and possibly determining the input–output relationship between the abundance of the pests and the level of damage. An evaluation requires use of at least one of experimental design, sampling and analysis. These are described in sequence in this section.

Vertebrate pest damage can be estimated qualitatively or quantitatively. Usher *et al.* (1988) qualitatively ranked the impacts of invasive pests in nature reserves. A severe impact was either a proven or potential extinction of a native population, or a significant effect on ecological processes. A moderate impact was a significant change in long-term abundance of native species, and a slight impact was any other lesser effect.

2.3.1 *Experimental design*

To evaluate damage (this chapter) and pest control (Chapter 3) a distinction needs to be made between experimental design and sampling schemes. The phrase ‘sampling design’ is an ambiguous mix of the two, so should be avoided.

Experimental design is the spatial and/or temporal arrangement of factors and treatments. Treatments are what we do, the procedures whose effects are to be measured and compared. The treatments consist of all combinations that can be formed from the different factors (Cochran & Cox, 1957). Factors may be pest abundance, crop species or varieties and distance from roosting sites. Each factor may have two or more treatments. For example, if crop variety is a factor and four crop varieties are tested, then there are four treatments. If a pest population is culled in three different ways then the culling is a factor with three treatments. The several treatments within a factor are also called levels. Each treatment may be replicated and hopefully is. Sampling, also called sampling scheme, is the spatial and/or temporal arrangement of sampled units. Such sampled units may include, for example, quadrats in a crop, days in a year.

Examples of five experimental designs are shown in Fig. 2.2. These and other designs are discussed further in Cochran & Cox (1957), Steel & Torrie (1960),

Fig. 2.2. Diagrams of five experimental designs. The boxes represent crops used in a hypothetical study of the effect of crop variety (1 to 3) on damage by rodents. Rodents are prevented from moving between fields by barriers. The designs are discussed in the text. There are two treatment replicates (three occurrences) of each variety in each design, except for balanced incomplete blocks where there is one replicate (two occurrences).

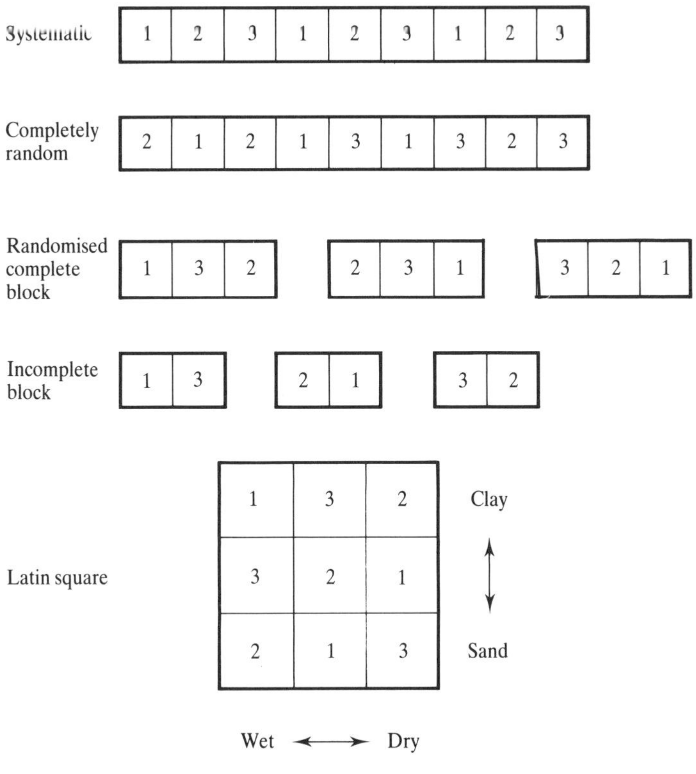

Snedecor & Cochran (1967), Krebs (1989) and Skalski & Robson (1992). Many of the designs were originally described in the 1920s and 1930s, as is evident from these texts and Preece (1990). Maindonald (1992) argues convincingly for authors of papers in applied biological research, which would include vertebrate pest research, to give much clearer descriptions of experimental layout, give reasons for choice of design and take care in analysing data. In the hypothetical study of the effect of crop variety on damage by rodents shown in Fig. 2.2, there is one factor (crop variety) with three treatments (three varieties). In a systematic design the treatments are allocated in a predictable manner, for example, crop variety 1 is always next to crop variety 2. There are two replicates (three examples) of each variety. In all the other designs there is some element of randomisation in the allocation of treatments. A completely randomised design occurs when each treatment and its replicates are arranged at random. There are no restrictions on the arrangement. A completely randomised design can have several factors each at several levels, for example, two factors each at two levels for a total of four treatment combinations, in which case it is called a factorial design.

In some designs there are more treatment combinations than sites. Then each site will not include a complete set of treatment combinations. An incomplete block design is then used. In Fig. 2.2 there is one replicate of each treatment. If each block includes a full set of treatments then it is a complete block design. The arrangement of treatments is random within each block, and re-randomised between blocks. It is extremely unlikely that the actual arrangement of treatments will be identical in each block.

When there are gradients across a site, for example, soil type and moisture, then treatments can be restricted so that each segment of the gradient has a full set of treatments. Such a design is termed a Latin square design. In it each row and column in the experimental area has an example of each treatment. Incomplete block and Latin square designs have real practical advantages but some statistical limitations. A practical advantage of a balanced incomplete design is that it can be used when only a small number of plots are available. The practical advantage of the Latin square design is that it recognises features of the site in the design, and so should reduce experimental error (variation). In a Latin square design the interaction of two treatments cannot be tested (Snedecor & Cochran, 1967).

Split-plot designs are used when an interaction between treatments is of particular interest and one treatment is of more interest than the other. The treatment of greater interest is used as the sub-plot treatment and the treatment of less interest as the main-plot treatment. The key feature of a split-plot design is that the main-plot treatment is applied over the same plots as the sub-plot treatments (Snedecor & Cochran, 1967). The sub-plot treatments are wholly contained within each level of the main-plot treatment. The split-plot design is not illustrated in Fig. 2.2 as one would not usually plant one crop variety wholly within a field used for planting another crop variety.

Confounding of treatments should be avoided. Confounding occurs when the effects of two or more treatments cannot be distinguished from one another. For example, if a study of the effect of crop variety on damage by rodents were to use one crop variety in one year but a different crop variety in another year then the effects of variety could not be separated from the effects of years.

Experimental controls, such as sites with no pests, may or may not be necessary. Cochran & Cox (1957) considered that when the effect of a treatment is unknown or is variable then an experimental control is needed. An example here would be when the damage by a species is unknown. When the effect of a treatment is well known then the experimental control is not as necessary. Hence, in the early stages of an investigation into a suspected vertebrate pest, an experimental control, wherein the pest is absent, is needed. As the study proceeds and the cause–effect relationships between the pest and the damage are established then impacts may be assessed with less use of experimental controls; however, more often than not, I would recommend their use.

Hairston (1989) and Krebs (1989) considered that experimental controls were necessary in ecological studies. If such studies are exploratory then that view is consistent with that stated by Cochran & Cox (1957), a book not referred to by Hairston (1989). Krebs (1989) recommended that experimental controls should be contemporaneous with the treatment; that is, they should be monitored at the same time, rather than simply before and after (sequential) use of treatments. This aspect is discussed further in section 3.4. When in doubt, use replicated experimental controls, and review the work of Eberhardt & Thomas (1991) and Skalski & Robson (1992), and references therein, on the limited power of statistical tests which have few treatment replicates.

The experimental designs shown in Fig. 2.2 suit standard statistical analyses, such as analysis of variance. These can be used to estimate the effects of crop varieties on rodent damage or other topics. The null hypothesis is that there is no difference in the level of damage between varieties. The designs have no experimental control, such as crops which have no rodents. If not much is known about rodent damage then an experimental control should be included. If by chance or otherwise, some crops have more rodents than others, then the estimated damage may be partly related to crop variety and partly to rodent abundance. Two approaches could be used to investigate this and to disentangle the two likely sources of variation. Both approaches require estimates of rodent abundance, either true abundance or a linear index of abundance. The latter is described by Caughley (1980).

The first approach is to test specifically the relationship between pest abundance and damage. Such a test is by regression or correlation, where abundance is the independent variable (x) and damage is the dependent variable (y). The second approach is to use pest abundance as a covariate in the analysis. The effects of rodent abundance are then stabilised, so the effects of abundance are effectively removed and the effects of crop varieties can then be tested. The analysis is then called analysis of covariance (Snedecor & Cochran, 1967). These two approaches are not independent. The analysis of covariance assumes that a relationship between pest abundance and damage exists, and that the relationship is linear. Table 2.1 lists results of such linear analyses. Some studies have used correlation analysis to study the pest–damage relationship. Correlation does not necessarily establish cause and effect relationships.

The relationship between the number of pests and the level of damage is an extension of the concept of an experimental control. It is variously called the input–output or dose–response relationship (Dillon, 1977) and is part of the modelling work described in section 5.4. The experimental control is simply where the number of pests, and the damage, are both zero.

If interest is mainly in a relationship, such as the regression between pest abundance and damage, then it is important to consider whether to have more treatment levels or more replicates at each level. In general, response relationships are better estimated by having more levels and fewer replicates at each level, especially if the relationship is curved (Dillon, 1977). In either case there is a strong need to have treatment and sample replicates because of natural variability. This is discussed further in section 3.4.

Use of conventional experimental designs allows stronger inferences and conclusions about effects of vertebrate pests than indirect inferences from unreplicated non-treatment studies. However, sometimes the designs can be difficult to use because pests move between sites. Fencing or netting can be used in some studies, or sites need to be located far apart to overcome movement.

2.3.2 *Sampling*

Obtaining information about spatial or temporal variation usually involves sampling of populations. Studies of damage have often not reported sample size, how many samples were adequate, or the merits of alternative sampling strategies.

There are many types of sampling scheme. Steel & Torrie (1960), Cochran

Table 2.1. *Summary of studies of the linear relationship between pest damage and abundance. Where I have tested the relationship using data from a study, the source is listed as 'After ...'*

Damage	Pest	Relationship	Significance	Source
Aircraft strikes	Birds	Positive	$P<0.01$	After van Tets (1969)
Prevalence of tuberculosis	Badger	Positive	$P<0.01$	After Cheeseman *et al.* (1981)
Wheat	Lesser bandicoot rat	Positive	$P<0.01$	Poche *et al.* (1982)
Crop damage	Wild boar	Positive	$P<0.01$	Gorynska (1981)
Ground rooting	Feral pig	Positive	$P<0.05$	After Ralph & Maxwell (1984)
Pistachio nuts	Crow	Positive	$P<0.05$	Crabb, Salmon & Marsh (1986)
Pistachio nuts	Scrub jay	Positive	$P<0.05$	Crabb *et al.* (1986)
Changes in ground rooting	Feral pig	Positive	$P<0.05$	Hone (1988a)
Ground rooting	Feral pig	Positive	$P<0.01$	Hone (1988b)
Sugar cane	Rat	Positive	$P<0.01$	Lefebvre *et al.* (1989)
Pasture composition	Rabbit	Positive	$P<0.05$	Croft (1990)
Wool production	Rabbit	Negative	$P<0.05$	Croft (1990)
Wheat	Brent geese	Positive	$P<0.05$	Summers (1990)
Crop damage	Wild boar	None	NS	Mackin (1970)
Crop damage	Wild boar	None	NS	Andrzejewski & Jezierski (1978)
Ground rooting	Feral pig	None	NS	After Cooray & Mueller-Dombois (1981)
Barley	Ducks	None	NS	Gillespie (1985)
Aircraft strikes	Gulls	None	NS	Burger (1985)
Ground rooting	Feral pig	None	NS	Hone (1988a)
Prevalence of tuberculosis	Badger	None	NS	After Cheeseman, Wilesmith & Stuart (1989)
Oilseed rape	Brent geese	None	NS	McKay *et al.* (1993)

(1977), Caughley (1980) and Krebs (1989) give detailed descriptions and Rennison & Buckle (1988) apply the schemes to estimating rodent impacts on crops. Common schemes used in vertebrate pest research include authoritative, systematic, random and stratified random sampling.

Authoritative sampling occurs when the experimenter, the authority, decides where and when sampling will occur. No consideration is given to statistical issues or prior information but usually it is accompanied by the statement, 'this looks like a good spot'. In *systematic sampling*, sample units are arranged regularly; for example, there is a constant distance between them, such as plots within a transect. The transects may also be a regular distance apart. In this example both the plots within transects and the transects are arranged systematically. There is no random selection of units. The main concern is to avoid harmonic patterns in the environment. In each of the other sampling schemes (random and stratified random) there is at least one level of randomisation.

If the units to be sampled are selected completely at random, then that is simple *random sampling* (also called completely random sampling). Combinations of systematic and random sampling are used, for example, where plots are placed regularly within transects when the transects have a starting point that is randomly selected. The sampled units can be divided into groups of similar levels of damage, and the groups called strata. *Stratified random sampling* involves selecting units at random within each strata. The randomisation is independent between strata. Cochran (1977) reported that stratified sampling gives greater precision (lower residual variance or standard error) than simple random sampling. To estimate the extent of impacts, for example predation, the process of stratification requires prior knowledge of the extent of predation in different areas so the areas can be allocated to different strata. Post-sampling stratification which uses the variable sampled during a survey can give biased results (Caughley, 1977).

When impacts are rare or difficult to locate then *survey sampling* methods can still be used. Sudman, Sirken & Cowan (1988) described efficient sampling methods for geographically clustered sites or populations and for rare and dispersed populations. An example of clustered populations could be sheep flocks bordering forests. Cluster sampling could be used in the former case (populations clustered), though the sampling variance may be biased compared to simple random sampling. In the latter case (dispersed, rare populations) *network sampling* may be an efficient method. In network sampling of predation, sites (ranches or farms) are sampled at random and respondents are asked about other sites in the area which are believed to suffer predation. Sites so identified are then surveyed.

If the aim of a study is to classify damage as high, medium or low, rather than estimate the actual level, then *sequential sampling* could be used. The loss of accuracy in describing the level of loss is traded off against the reduction in field time to obtain a correct classification. Waters (1955) and Ruesink & Kogan (1982) described the method for surveys of insect pests and Krebs (1989) for other work. Three sets of information are necessary to use sequential sampling. Firstly, the frequency distribution of pests, or levels of damage, per sampled unit must be determined. The distribution may be binomial, normal, Poisson or negative binomial. Secondly, enough must be known about the pest or levels of damage to set limits to previously determined classes; that is, how much damage or how many pests correspond to a low or high level. Thirdly, the probabilities of correctly classifying populations into a pest abundance or damage class must be specified.

The results of previous studies or a preliminary study can be used to estimate the sample size needed to obtain a specified level of precision or to be able to detect a specified difference between treatment means (Cochran, 1977; Eberhardt, 1976, 1978). Use of existing data is encouraged as it can greatly increase precision and cost-effectiveness. The final choice will be a trade-off between statistics and practicalities. The more aware you are of the trade-off and the consequences, the more efficient the study will be. Dale *et al.* (1991) showed by computer simulation that the frequency distribution of the data influenced the estimated mean and precision of samples, especially when sample sizes were small. The sample means of data from a log-normal distribution were biased compared with the means from a normal distribution. The difference lessened as sample sizes increased. Similarly, the precision of sample means from a log-normal distribution were lower (higher standard error to mean ratio) than for a normal distribution. Given that non-normal frequency distributions occur (see section 2.2) these results should be considered when estimating sample sizes.

Finally, a distinction must be made between a treatment replicate, an issue of experimental design, and a sample replicate, an issue of sampling schemes. Many studies refer to replicates but do not distinguish between the two types. Treatment replicates are what the name suggests – replicates of the experimental treatment being studied, for example, crop variety. Sample replicates are multiple examples of the sampled units, for example, quadrats within each crop variety. Both treatment and sample replicates must be independent, or if not, must be analysed as not being independent. Hurlbert (1984) discusses the difference further and the problems of confusing them.

2.3.3 *Planning and analysis*

In the final planning of damage, and control evaluation, one should consider increasing precision to get a better estimate of the effect of treatments being studied. Options for increasing precision are listed in Table 2.2 in a suggested order of use. To get better estimates of damage it is important to increase precision. This is the other side of the coin of making incorrect conclusions from studies, as encapsulated in type I and type II errors (concluding that differences occur when they do not, and concluding that differences do not occur when they do, respectively).

The probability that a statistical test will detect a difference when it actually occurs is called the power of the test, and is calculated as one minus the probability of a type II error. Whysong & Brady (1987) showed that a practical way of increasing the probability of detecting changes in vegetation when sampling, using frequency of occurrence data, was to increase sample size. Detecting changes of plus or minus 10% may require over 500 sampled units, while detecting changes of plus or minus 5% may require over 1000 sampled units. Gerrodette (1987) discusses similar ideas for detecting rates of increase of populations.

Research often concentrates on the response of damage or pest populations to a level of control rather than the relationship between the level of control and the response to control. For example, a study may compare crop yield at sites with birds with crop yield at sites with no birds (analysis of difference) rather than crop yield at sites with different numbers of birds (analysis of response) (Fig. 2.3). The differences between this analysis of response and the analysis of difference were discussed by Dillon (1977) for agricultural research. The analysis of response often occurs in vertebrate pest research only in laboratory

Table 2.2. *Checklist of methods for increasing precision in experimental design and sampling when estimating damage of vertebrate pests. The same methods can be used for studies of vertebrate pest control*

1. Do a preliminary survey to estimate sample size
2. Stratify experimental units into homogeneous units, that is, make them similar or use local control
3. Refine methods
4. Use an efficient experimental design (such as a complete block or factorial design rather than a completely randomised design)
5. Sample without replacement
6. Increase sample size
7. Measure a covariate and use analysis of covariance

or pen experiments, sometimes for the real practical reasons that large field studies are difficult to conduct.

If the analysis of damage tests the difference between, for example, pests and no pests, then an efficient strategy is to increase the number of treatment replicates. If, however, the emphasis is to estimate the relationship between the level of pests and a response, such as damage, then the primary concern is the number of levels of pests not the number of replicates. Similarly, this applies to studies of the relationship between pest damage and the level of control. Dillon (1977) advanced a similar argument in the analysis of response in agricultural research, where he highlighted the trade-off between treatment levels and treatment replicates. In general, the researcher should increase the number of levels and decrease the replicates at each level (Fig. 2.3) if resources limit the total number of study sites (which of course they invariably do). This will be particularly relevant if the relationship to be estimated is a curve (Dillon, 1977). The point is also very relevant to the analysis of response of pest abundance to control as described in Chapter 3.

The selection of which design, sampling scheme and analysis to use is best done in consultation with a biometrician (Eberhardt, 1976). Biologists repeatedly assume they understand experimental design and sampling only to

Fig. 2.3. Hypothetical results of studies between crop yield and pest abundance. (*a*) Yield data from six crops, three where there were no pests and three where there were 100 pests. Analysis of such data would be a test of difference. (*b*) and (*c*) Yield data from six crops but with more levels of pest abundance than shown in (*a*). Analysis of such data would be a test of relationship or analysis of response. The figures show the trade-off between the number of levels of number of pests (2, 3 or 6 respectively) and the number of replicates for each level (2, 1 or 0 respectively).

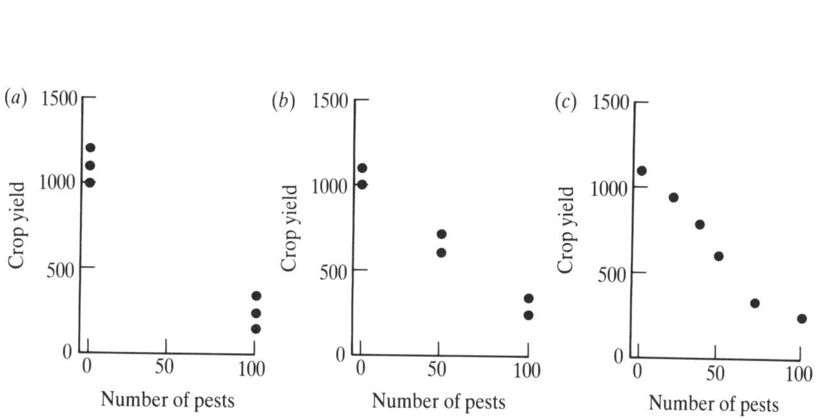

be told by a biometrician after data collection that they did not understand. The selection is usually a trade-off between statistical elegance and practicalities. Most statistical textbooks discuss this, as does section 3.4.

One final topic in the analysis of pest damage is noted. In studies of the effects of vertebrate pests on plants, there may be compensatory growth by the plants that complicates the interpretation of results. Belsky (1986) reviewed the topic of compensatory growth as a response to herbivory. It was suggested that when the total dry weight of plant tissue in control plots was less than that of grazed or clipped plants, then it should be called overcompensation. Undercompensation occurred when the dry weight of grazed plants was less than that of control plants. Such grazing or clipping causes damage when dry weight is less than would be expected (presumably less than the difference in the amount removed).

Judenko (1973) described a method for estimating the effect on crop yield of compensatory growth after attack by a vertebrate pest. Data are required on the yield of attacked plants, adjacent unattacked plants and non-adjacent unattacked plants. The two sets of 'control' plants are used to estimate different effects. The non-adjacent unattacked plants are used as true controls. The adjacent unattacked plants are used to estimate any compensatory increase in yield resulting from the impacts. Poche *et al.* (1982) reported compensatory growth of wheat to be important in reducing rodent damage in parts of Bangladesh, and Cummings *et al.* (1989) reported that compensatory growth may reduce bird damage to sunflowers in parts of the U.S.A.

A range of studies of vertebrate pest damage are now described. Several conclusions are then made at the end of the chapter.

2.4 Predation of livestock

2.4.1 *Problem definition and species*

Predation is defined as an interaction between individuals of different species where one (the predator) kills and eats the other (the prey). This section will examine predation of sheep (*Ovis aries*) by coyotes (*Canis latrans*), dogs (*Canis familiaris*), feral pigs (*Sus scrofa*), foxes (*Vulpes vulpes*), and several bird species particularly crows and ravens (*Corvus* spp.) and eagles.

2.4.2 *Methods of study*

Predation by vertebrate pests has been studied by many methods. Seven are described here, from the most general to the more specific.

Questionnaires (method 1) involve sending, usually mailing, a questionnaire to participants who may be selected randomly from a list of producers, or selected in some other way. If the aim of a study is to infer results from the sample of participants to other producers in the area, then a random sample is obligatory. Precision may be increased by using stratified, two-stage or cluster sampling (Cochran, 1977). Questionnaires have the advantage of being relatively easy to organise and cheap, though care is needed to ask the right questions. An important limitation is that not all those sent questionnaires respond. Schaefer, Andrews & Dinsmore (1981) reported that in Iowa only 39% of mailed questionnaires were returned. The low rate of return can be improved by sending a follow-up questionnaire to those who did not respond.

Postcard surveys (method 2) involve sending a postcard to producers on which they submit periodic estimates of predation losses. Schaefer *et al.* (1981) reported that in Iowa not all producers returned cards each month (mean 69 of 110 = 63%), and the number declined significantly over time. This is to be expected as producers are probably less likely to submit a card when they consider no predation has occurred. Hence, the method could underestimate the absence of predation (overestimate predation) as producers are likely to lose interest in a project over time. The loss of interest could be reduced by regular feedback of research results to producers, and at the start of the survey clearly defining an endpoint. Robel *et al.* (1981) used the postcard survey method in Kansas, but reduced the non-return rate by sending producers a reminder when they did not submit cards, visiting them and conducting a personal interview.

In personal interviews (method 3) producers are interviewed, after some selection process. Dorrance & Roy (1976) used this method in Alberta, where producers were selected systematically by choosing every twentieth name on a membership list. This method can be expensive because the interviewer needs to travel to farms or ranches, and interview people personally. The interviewer should decide prior to each interview to record it, take notes or have participants complete a survey form.

Domestic animal claims (method 4) are made by producers submitting claims for compensation. Verification of predation can be required, by witnesses, and claims submitted within several days of the kill. Schaefer *et al.* (1981) used the method in Iowa to study predation by coyotes and dogs.

Field post-mortem examinations (method 5) use detailed examination of dead livestock to identify the species of predator, based on the distinctive pattern of the kill (Rowley, 1970; Wade & Bowns, 1982; Jordan & Le Feuvre, 1989). Blood on the carcass indicates it was killed rather than being eaten after death. Extensive skinning of carcasses can be used to obtain evidence of predation, such as haemorrhagic patches around wound sites. An alternative approach is to examine the dead specimens of the suspected predator. Croft &

Hone (1978) used this approach when examining the stomach contents of foxes in New South Wales. The results can be ambiguous, however. Sheep remains in the stomach of a fox do not distinguish between the fox having eaten a dead sheep or having killed a sheep, most likely a lamb, and eaten some of the carcass. The method could be refined by looking for maggots or other signs of prey decomposition prior to ingestion, or by labelling prey with radio-transmitters which signal death but still allow location of the transmitter.

Direct observation of predation (method 6) can be difficult, especially of nocturnal predators. This can be overcome by using night vision equipment if the predators or prey will allow observation without disturbing them. Radio-tracking equipment can also assist in locating prey and predators. Rowley (1970) used direct observation to study crows and ravens in Australia and concluded most species did not and could not kill lambs because of the shape of the birds' bills. A combination of direct observation and field autopsies was used by Gluesing, Balph & Knowlton (1980) in Montana to study predation of sheep by coyotes. Pavlov & Hone (1982) used direct observation of feral pigs in lambing flocks in New South Wales to confirm that predation occurred and to identify the pigs that were killing lambs.

The experimental method (number 7) estimates the effect of predation as the difference between prey abundance in areas with predators and areas without predators. Plant *et al.* (1978) and Pavlov *et al.* (1981) used the experimental method in New South Wales to estimate the effect of feral pigs on lamb production. The method can be elaborated to estimate the relationship between prey abundance and the number of prey killed (the functional response to predator–prey theory; Krebs, 1985 and further discussed in section 5.3).

The various methods have their strengths and weaknesses. Schaefer *et al.* (1981) reported that questionnaire and postcard surveys gave similar results, except that a higher percentage of predation was attributed to coyotes in the questionnaire survey. This could, however, have been confounded with area effects, as the two study sites differed. They also reported differences in results between the questionnaire survey and domestic animal claims, and between the postcard survey and domestic animal claims. The accuracy of claims was also investigated by Schaefer *et al.* (1981) who reported that some producers made invalid claims, especially those with least experience in identifying predation. The personal interview may have similar biases to the questionnaire. A potential bias of the experimental method is that predator exclusion from one area may cause increased predation in another area. This could occur if predators are pushed out of the experimental control area and into an adjacent treatment area. When two or more predators occur in an area the experimental method may show no effect of control of one predator species if substitution of predators occurs.

2.4.3 *Analysis*

The extent of predation can be estimated as the number, or percentage, of prey killed. More detailed analysis can examine the spatial and temporal frequency distributions of predation, and the sources of that variation. The latter may be developed into a relationship to predict the future levels, locations and times of predation.

Wagner (1972, 1975), Wagner & Pattison (1973), Dorrance & Roy (1976), Sterner & Shumake (1978) and Robel *et al.* (1981) concluded that the frequency distribution of spatial variation in losses to predators in northern America, or parts thereof, took the form of a Poisson or Poisson-type distribution. Boggess, Andrews & Bishop (1978) concluded that the distribution showed a logarithmic decline. Most ranchers sustained light losses but a small number experienced heavy depredations. None of the studies, however, tested the fit of the data to a Poisson distribution. My analysis of data in Robel *et al.* (1981) is now described. The data were the spatial frequency distribution of damage.

The Poisson distribution describes the expected number of occurrences of events occurring at random. The probability of an event, r, is given by:

$$P(r) = (u^r \mathrm{e}^{-u})/\, r! \qquad (2.1)$$

where u is the mean occurrence, r is the number of occurrences of predation, $r!$ is r factorial $(= r \times (r-1) \times (r-2) \ldots \times 1)$ and e is the base of natural logarithms (2.71828). The Poisson distribution has the unique properties that the mean equals the variance and the mean expectation of events is constant from trial to trial (Snedecor & Cochran, 1967). The shape of the Poisson distribution varies from a hyperbolic to a normal-like curve (Moroney, 1965). The hypothesis that the set of predation data was described by a Poisson distribution was examined by a Chi-square goodness of fit test (Snedecor & Cochran, 1967). Probabilities for each class were estimated for each level of predation within the class, then summed over those levels.

The reported levels of predation were significantly different from a Poisson distribution (Table 2.3). The number of producers that had no sheep killed by predators was much higher than predicted by a Poisson distribution. Thus, predation was not a random event. The frequency distribution was also significantly different from a negative binomial distribution (Table 2.3). The fit of the data to a negative exponential function ($y = a\mathrm{e}^{-bx}$) was also estimated. The x variable was the mid-point of each class of the level of predation. The fitted regression ($y = 35.73\,\mathrm{e}^{-0.12x}$) was significant ($r = -0.89$, df $= 3$, $P < 0.05$).

The tabulated expected values (Table 2.3) were similar to the observed values, except for the underestimate (50 observed and 36 expected) of the number of ranches with no sheep losses.

The fit, or lack of fit, of data to any particular frequency distribution should not be over-interpreted. Lack of fit to a Poisson distribution suggests predation does not occur at random, which is not surprising. The point of this analysis is to demonstrate how a frequency distribution can be tested for goodness of fit, a step beyond assuming a particular frequency distribution occurs.

An example of temporal variation in sheep predation in the U.S.A. was described by Terrill (1986). The estimated losses between 1940 to 1985 inclusive varied with the frequency distribution of losses (Fig. 2.4) being positively skewed. The sources of this variation in sheep predation should now be examined. A pest–damage relationship has been hinted at in the literature. Wagner (1972), after examining data on coyote abundance and predation losses, considered that there was a (positive) correlation between coyote population density and the level of sheep loss, but did not test the association.

Variation in sheep predation was associated with differences in husbandry methods in Kansas (Robel *et al.*, 1981). Night confinement, farm dog ownership, lighting in corrals, distance to nearest town, fall (autumn) lambing and flock size contributed significantly to variation in predation rates by coyotes and dogs. Differences in biophysical attributes, such as topography and vegetation type also contributed to variation in predation by coyotes and

Table 2.3. *A comparison of the observed and expected extent of sheep predation. Observed data are from Robel* et al. (1981). *Expected data are for Poisson, negative binomial or exponential distributions. Tabulated expected numbers have been rounded to the nearest integer*

		Expected		
Number of sheep killed/producer	Observed	Poisson	Negative binomial	Negative exponential
0	50	5	34	36
1–4	29	79	45	27
5–8	9	18	14	16
9–12	10	<1	6	10
13–20	7	<1	2	5
Chi-square		25624.3	20.9	
df		3	2	
Significance		$P<0.005$	$P<0.005$	

dogs. Each management factor was examined separately by a Chi-square test, so a limitation of the analysis, as noted by the authors, was that any interaction between factors could not be determined. Similarly, the cumulative proportion of variation in predation due to those factors was not estimated. The percentage of sheep killed decreased as flock size increased (Robel *et al.*, 1981).

Pavlov *et al.* (1981) reported in their Table 2 that the number of lambs killed by feral pigs increased as lamb abundance increased. These results are expected from the functional response of predator–prey theory, which links prey captures to prey abundance (Holling, 1959; Krebs, 1985; Begon, Harper & Townsend, 1990). My analysis of their data fitted three functional response relationships. A linear relationship was estimated by least squares regression, firstly without forcing the line through the origin and secondly by forcing the line through the origin and assuming that the standard deviation of the dependent variable increased as the level of the independent variable increased. In the third analysis a curvilinear relationship was estimated by iterative least squares.

The slope of the unforced linear regression (0.28) was significantly different from zero ($t=4.67$, df$=4$, $P<0.01$), but the slope of the forced linear regression (0.69) was not significantly different from zero ($t=1.30$, df$=4$, $P>0.05$). The curvilinear equation was of the form $k=a(1-e^{-bL})$ where k is the number of lambs killed per two weeks, a is the maximum (satiated) intake of lambs, L is the estimated number of lambs born per two weeks and b is a measure of foraging efficiency (higher values of b correspond to more rapid eating of the

Fig. 2.4. The frequency distribution of sheep losses to predators in the U.S.A. (After Terrill, 1986).

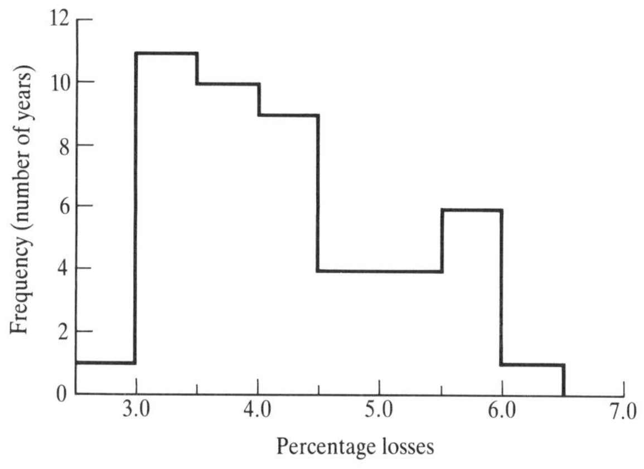

food source). The equation was estimated by rearranging the equation to a regression between the number of new lambs (L) as the dependent variable, and using $\ln(1-(k/a))$ as the independent variable. The regression was forced through the origin, assuming that, for increasing levels of the independent variable, the standard deviation of the dependent variable increased (Snedecor & Cochran, 1967). The slope ($-1/b$) of the fitted curvilinear regression was significantly different from zero ($t=9.70$, df=4, $P<0.001$). Overall, the curvilinear regression had a better fit to the data, as it had a higher t value for the test that the slope was zero.

The two significant fitted regressions (unforced linear and curvilinear) had different predictions (Fig. 2.5). At low and high abundance of lambs, the unforced linear regression predicted a higher number of lambs killed, but at intermediate numbers of lambs, the curvilinear regression had the higher prediction. The predictions of each fitted regression could be tested with an independent set of data, though this has not been done.

Pavlov *et al.* (1981) noted a limitation in the original data which was that if the predators (pigs) kill only one of a set of twin lambs, then an analysis of the effect of predators on the percentage of ewes lactating (that is, with young) would be biased. Obviously, lambs would have been killed but the percentage of ewes lactating would not have changed. Also, predation of a twin lamb may result in a compensatory decrease in the likelihood of death of the surviving twin.

Pavlov & Hone (1982) estimated the hunting success of feral pigs which chased lambs. Prey were caught in 10 of 42 chases (23.8%). The hunting success

Fig. 2.5. The estimated relationships between the number of lambs killed by feral pigs (k) and lamb abundance (L). Data are from Pavlov *et al.* (1981). Equations are discussed in the text.

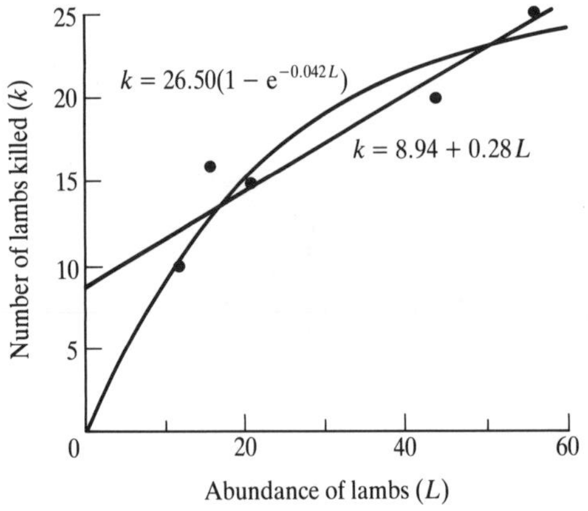

or probability of capture was 0.238, so the probability of not catching a lamb was 0.762. The number of lambs that a feral pig has to chase to be 99% certain of catching a lamb can be estimated from the hunting success. The random variable, the number of chases until the first success, has a geometric distribution (Grossman & Turner, 1974). The geometric distribution is similar to the binomial distribution in that both are concerned with repeated experiments with only two outcomes, in this case success or failure of a chase. When the number of chases is fixed, then the number of successes has a binomial distribution. When chases occur until the first success, then the number of unsuccessful chases has a geometric distribution. In this study a feral pig must chase n lambs, where n is chosen so that $1-0.762^n<0.01$ ($=1-0.99$). Hence, a feral pig must stalk at least 17 lambs to be 99% assured of catching prey.

The statistical analysis of control of predation is addressed in section 3.15, economic analysis in section 4.4 and modelling in sections 5.5 and 6.7.

2.5 Infectious diseases

2.5.1 *Problem definition*

Many wild vertebrates are known or suspected to be infected with diseases of humans or livestock. As a consequence these wild vertebrates are often described as pests without closer scientific study. In some situations infectious diseases of endangered species may be of interest or concern (Scott, 1988). This can occur when the population is very small, has a homogeneous genotype, is confined to a small area or held in captivity. The topic of diseases, especially human diseases, can evoke great passions and pressures to obtain quick and dramatic effects in pest control. The topic does, however, require careful study. Many people's reaction to the topic of diseases and disease control may be linked to fears of events like the spread of black plague, and to definition of the word pest (Cherrett *et al.*, 1971), as described in Chapter 1. The importance of diseases is highlighted in Macdonald (1980), which states that rabies may cause 15 000 human deaths annually worldwide. However, not all of these deaths originate from people acquiring the infection from wildlife.

The main issues in diseases and vertebrate pests are to answer the following questions: 'What is the role of the wildlife as hosts of the disease?'; and 'What are the effects of infection on human health or livestock production or conservation of native species?' These questions are easier to ask than to answer and have not been answered satisfactorily to date.

2.5.2 *Types of disease*

Infectious diseases are caused by pathogens that have been classified as microparasites (viruses, bacteria and protozoans) and macroparasites (helminths and arthropods) (Anderson & May, 1979). Fungi were not listed by Anderson & May (1979) but would be classified as microparasites. The diseases of most interest to us are those caused by viruses, such as rabies, foot and mouth disease and classical swine fever, and diseases caused by bacteria, such as tuberculosis and bubonic plague. Many of the infectious diseases are transmitted directly between infected and susceptible hosts but others are transmitted from infected hosts via vectors or alternative hosts. In different situations humans or livestock may be the alternative hosts, while biting insects are often the vectors.

2.5.3 *Methods of study*

Lilienfeld & Lilienfeld (1980) described two steps in epidemiological research. Firstly, determine a statistical association between a feature of the host or environment and a disease, and secondly, make biological inferences from that association. The data listed in Table 2.4 are concerned with the first step. Diagnosis (point 1) simply identifies the disease is present. The need for careful diagnosis is illustrated by the apparent misdiagnosis of rabies in Hawaii in 1967 (Tomich, 1986). Hawaii was previously rabies-free and the apparent case caused considerable concern in the community. Proof of transmission to other species (point 2) establishes that wildlife can be an alternative host and hence could contribute to reduced agricultural production or some other measure of damage. Proof of transmission within the wildlife population (point 3) suggests that the disease may establish in the population, and persistence (point 4) suggests that transmission occurs between individuals or between individuals via the environment and that in the environment the pathogen is long-lived. A significant relationship between wildlife abundance and disease prevalence, or incidence (point 5) is useful for analysing the potential response

Table 2.4. *Checklist of data needed to establish the role of wildlife hosts in disease maintenance or spread to humans or other animals*

1. Pathogen or disease is present in the population
2. Pathogen is transmitted to humans or other animals
3. Pathogen is transmitted between individuals in the population when alternative hosts, such as humans or livestock, are absent
4. Pathogen persists in the population beyond the maximum longevity of any individual host
5. Significant relationship exists between wildlife abundance and equilibrium or non-equilibrium prevalence, or incidence
6. Basic reproductive rate of disease in wildlife is greater than or equal to one

of disease prevalence, or incidence, to pest control. For example, a significant positive relationship suggests that disease prevalence, or incidence, could be reduced by reducing pest abundance. This issue is further discussed in section 5.6 as is the basic reproductive rate (point 6) listed in Table 2.4. A basic reproductive rate of greater than one means the disease is spreading in the wildlife population. This is another way of describing point 3 in Table 2.4. If the basic reproductive rate is less than one then the disease will die out, so the wildlife hosts will not be important in disease.

There are three common methods for studying infectious diseases. Surveys (method 1) identify the disease in hosts or signs of present or past infection such as antibodies. The hosts can be selected at random, or all hosts can be surveyed systematically. Such studies are often retrospective, describing the results of infection some time in the past. Macdonald (1980) described the results of many surveys of wild animals for rabies. Cheeseman *et al.* (1981) used this method to detect bovine tuberculosis in badgers (*Meles meles*) in south-western England, and Coleman (1988) used the method to detect bovine tuberculosis in possums (*Trichosurus vulpecula*) in New Zealand. Detection of a disease may require evaluation of alternative methods of diagnosis such as reported by Pritchard *et al.* (1986) for tuberculosis in badgers.

The distinction between surveys and experimental studies (method 2) in epidemiology is whether the investigator has control over the assignment of individuals to a study group or treatment (Lilienfeld & Lilienfeld, 1980). Only in experiments does the assignment occur. The ability of a pathogen to produce disease in a host, or for an infected host to transmit disease to other hosts of the same or different species, can be determined by experimental inoculation. Snowdon (1968) examined the susceptibility of a range of Australian wildlife to foot and mouth disease virus. Animals were experimentally dosed with the virus in a laboratory. A third method of studying infectious diseases is modelling. This is described in section 5.6.

The methods have limitations. The results of surveys are dependent on accurate diagnosis of the disease or pathogen. The experimental method can require extrapolating from the laboratory to the field situation. The limitations of modelling are described in Chapter 5.

2.5.4 *Analysis*

The analysis of disease data obviously depends on the questions being asked. A few preliminary analyses which estimate incidence and prevalence are often performed. Incidence of disease is the number of new cases of disease over a defined time period. Prevalence of the disease is the percentage of hosts which are infected at a defined time.

Detecting a disease requires sampling and the question of how many hosts to sample is addressed here. The discussion relates mainly to uncommon diseases in wildlife. The consequences of a low likelihood of finding a disease is explored in section 3.16. Sheail (1991) has an enlightening description of the detection, and attempted control, of myxomatosis in England in 1953. The first outbreak was reported up to three weeks after dead rabbits had apparently first been noticed.

The sample size needed to detect a disease is determined by the size of the population, the presence or absence of the disease (obviously), and the required reliability of the conclusions (Cannon & Roe, 1982). Essentially, the issue reduces to sampling until the first case of disease is detected. For example, in the case of disease in a host it is assumed that sampling is random without replacement so an individual animal cannot be sampled twice. The equation for the sample size was described by Cannon & Roe (1982) as:

$$\text{sample size} = (1-(1-c)^{1/d})(N-(d/2))+1 \qquad (2.2)$$

where N is the population size, d is the number of positives in the population (the number of hosts infected) and c is the desired confidence level, that is, the probability of finding at least one positive in the sample. Hence if $N=100$, $d=2$, and $c=0.99$, then the sample size needed to detect one case of disease is:

$$\text{sample size} = (1-(1-0.99)^{0.5})(100-(2/2))+1 = 90 \qquad (2.3)$$

As the number of positives (d) increases then the sample size decreases, as the population size (N) increases the sample size increases and as the reliability required (c) increases then the sample size also increases.

Often when studying wildlife populations the actual population size is unknown. The equation could be used as a guide by inserting a best guess for N in the equation. The estimated sample size is not very sensitive to differing values of N when N is large. For example, when $d=2$ and $c=0.99$, then the sample size for a population of 200 is 136, when $N=400$ the sample size is 174, and for $N=1000$ the sample size is 204. Similarly, if the number of positives is not known, which will invariably be the situation, then assume that it is low (maybe $d<5$ or 10) and use a larger sample size.

Bacon (1981) examined the consequences of unreported cases of rabies in British foxes assuming rabies had entered Britain. If the probability of detecting a rabid fox was only 0.02, as Bacon reported for Germany, and if authorities required 95% confidence in detecting rabies then it was estimated that 148 rabid foxes would occur in Britain before the first case was detected. The equation for estimating the number of cases (S) was:

$$S = \log_{10}(1 - P)/\log_{10}(1 - D) \tag{2.4}$$

where P is the probability of detecting at least one case and D is the probability of detecting an individual rabid fox. Hence using the previous example, $P = 0.95$ and $D = 0.02$, the estimated number of cases is:

$$S = \log_{10}(1 - 0.95)/\log_{10}(1 - 0.02) = 148 \tag{2.5}$$

For disease eradication with such inefficient surveillance (low D), a very efficient method of fox or rabies control is needed, or the efficiency of the surveillance scheme needs to be increased (that is, D needs to be increased).

Hone & Pech (1990) also used equation 2.4 to estimate the number of cases of foot and mouth disease that could occur in feral pigs in Australia should the disease enter the country. The reporting rate (D) was crudely estimated as 0.0015 and if the confidence level was 95% ($P = 0.95$) then 2002 cases could occur before the first case was detected. The estimates of Bacon (1981) and Hone & Pech (1990) can be combined with epidemiological models to predict the time delay until first detection and its implications for control. Those topics are discussed further in section 3.16.

The statistical analysis of the adverse effects of infectious disease on humans or domestic animals, caused by wildlife hosts, is limited. Emphasis has been on analysing disease parameters in the wildlife, rather than the essentially spill-over effects of disease into humans or domestic animals. The statistical analysis of infectious diseases is discussed in section 3.16, economic analysis in section 4.5 and modelling in sections 5.6 and 6.8.

2.6 Rodent damage

2.6.1 *Problem definition and species*

Rodents occur as agricultural, environmental, public health and household pests. The range of damage is well described in the early chapters of Prakash (1988). The species most commonly classed as pests are the black rat (*Rattus rattus*), Norway rat (*Rattus norvegicus*) and house mouse (*Mus domesticus* and *M. musculus*).

2.6.2 *Methods of study*

There are four common methods for studying rodent damage.

Questionnaire surveys (method 1) give estimates of damage to crops or other human activities and are carried out either as mail or face-to-face surveys. Ryan & Jones (1972) surveyed farmers by mail to estimate losses during a plague of house mouse in southern New South Wales. Differences in losses between types of crop and between areas were estimated. Poche *et al.* (1982) interviewed grain farmers in parts of Bangladesh using a questionnaire. The interviews aimed to obtain information on rodent species causing problems, types of damage and control methods used. Singleton *et al.* (1991a) mailed a questionnaire to all farmers who grew soybeans in parts of western New South Wales. Data on farm size and damage estimates were requested.

Field sampling (method 2) gives estimates of rodent damage in crops or other locations. A comprehensive description of sampling to estimate rodent damage to crops was given by Rennison & Buckle (1988). Advani & Mathur (1982), studying rodent damage to crops in Indian villages, selected plots at random along transects that were diagonals of crop fields. Different sized plots were studied but reported to give results that were not significantly different. The variable estimated was the percentage damage, measured as the number of plants or fruit damaged divided by the total number of plants or fruit examined.

The damage by lesser bandicoot rats (*Bandicota bengalensis*) in wheatfields in Bangladesh was estimated by randomly selecting villages within districts, and sampling four wheat fields for each farmer interviewed in each village (Poche *et al.*, 1982). Within each field quadrats were selected, though how was not stated, and rodent damage estimated. Additional measurements of wheat stem density and bandicoot rat burrows were obtained. Saunders & Robards (1983) used the method of Dolbeer (1975) to sample mouse damage to a sunflower crop in southern New South Wales. Individual flowering heads were selected and scored in classes of 0, 25, 50, 75 or 100% damage. Whether the heads were randomly or systematically selected was not specified.

Damage to palms was assessed by Wood & Liau (1984) in a Malaysian plantation. Fresh damage was that done on the previous two or three days, and overall damage occurred over the previous five months. A sample of two hundred palms was observed for each damage class, with the selection process appearing to be systematic, not random. Fresh damage was recorded as a percentage occurrence and the overall damage given a ranking score from zero for no damage to three for heavily damaged.

Lefebvre *et al.* (1989) surveyed damage by rats (*Rattus rattus*) in sugarcane crops in southern Florida. Transects in growing crops were systematically selected in the field edges and all plants on each transect were examined. The variable recorded was the number of damaged stalks per transect. Damage at harvest was estimated by stratifying each crop into an edge and centre area, and sampling separately in each. It was not specified whether the number of samples was proportional to damage or the variability in damage, though the latter gives higher precision. Cut stalks of sugarcane were selected at a number of randomly selected points in the edge and centre of each field and the number and proportion of stalks with damage recorded.

Field post-mortem examinations (method 3) can provide evidence of rodent-caused damage. Wood & Liau (1984) examined the stomach contents of *Rattus tiomanicus* in an oil palm plantation in Malaysia, which confirmed reported damage to palm oil fruit. If only a small proportion of rodents are doing the damage then a large sample size may be needed to detect confidently at least one positive result.

The experimental method (number 4) compares yields or estimates of damage between sites with and without rodents or with and without rodent control. Advani & Mathur (1982) used the latter method in crops in Indian villages. The effect of black-tailed prairie dogs (*Cynomys ludovicianus*) on pasture biomass in a part of South Dakota was estimated by comparison of plots with and without the prairie dogs (Collins, Workman & Uresk, 1984). As the study aimed to estimate the biomass of extra forage available to cattle, then ideally as noted by the authors, both sets of plots should have had cattle grazing the pastures.

Buckle *et al.* (1984) estimated damage to rice crops by rice field rat (*Rattus argentiventer*) in Malaysia by comparing crops with and without rodent control. Fields were selected randomly and groups of plant shoots (tillers) selected systematically within each field. The percentages of tillers which showed cuts by rats were estimated. Rice fields were sampled during the post-poisoning period only, so the level of damage before rodent control was not estimated.

The damage to crabapple (*Malus* spp.) trees by montane voles (*Microtus montanus*) and meadow voles (*Microtus pennsylvanicus*) in British Columbia was measured by Sullivan, Crump & Sullivan (1988a). The variable estimated was the proportion of trees girdled by the rodents. The exact method was not described but in each small plot presumably all trees were observed, rather than a sample of trees. This is obviously possible in small plots. Singleton *et al.* (1991a) sampled soybean crops in western New South Wales for mouse damage. Fields were sampled before and after poisoning by randomly selecting parallel transects in crops, then at pre-determined distances along each transect, counting the number of damaged and undamaged pods on a set number of plants. The number of damaged pods on the plants was used as an index of damage as some pods had previously been gnawed from plants.

The various methods have strengths and limitations, though a comparative evaluation appears not to have been published. The questionnaire survey method suffers from potential biases associated with those surveyed simply guessing the level of damage. The field sampling method should give unbiased estimates of percentage or actual damage if sampling is random and intensive.

However, many published reports do not give details of sampling methods to help others repeat the study or interpret results.

Barnett & Prakash (1976) discussed methods for direct field sampling and noted several sources of bias: hoarding of food, damage to grain bags and grain contamination. The field autopsy method allows greater confidence in interpreting results by showing whether the accused rodents have actually eaten the material of interest such as grain. The experimental method allows strong inferences about the effect of rodents on the level of damage by the comparison of sites with and without rodents, or at sites with more and less rodents. Ideally, the sites should be selected randomly or deliberately matched to be as similar as possible. Estimates of damage from data on per capita food intake and population abundance may be biased if animals hoard or cache food. A measure of rodent damage such as stem cutting may not reflect effects on grain yield. This is because the cutting may have a minor effect on plant physiological processes and yield or because the cutting occurs early in grain development and subsequently compensatory growth occurs.

2.6.3 *Analysis*

Several analyses have estimated the effects of rodents on agricultural production. Temporal trends in the frequency of cuts to wheat stems by lesser bandicoot rats in Bangladesh were plotted by Poche *et al.* (1982). Percentage damage was significantly positively correlated with stem density ($r=0.98$, $P<0.001$), the number of burrow entrances ($r=0.63$, $P<0.01$) and the number of burrow systems ($r=0.77$, $P<0.01$), and negatively correlated ($r=-0.95$, $P<0.01$) in an exponential manner to the distance outside the rats' burrow system. Mean damage was significantly less on irrigated (1.6%) than non-irrigated (9.4%) fields, as assessed by Student's t test. The bandicoot rats hoarded food down to 1.5 metres below the soil surface.

Advani & Mathur (1982) presented the data on percentage damage to crops and did not analyse them beyond simple descriptive statistics such as the mean and standard error. Their data on damage between years and between crops would be amenable to analysis of variance for the effects of years and crop types. Similarly, Wood & Liau (1984) simply graphed the temporal trends in percentage damage to oil palm plants. Saunders & Robards (1983) combined the data from field surveys and estimates of mouse population size, with reports from the literature of daily grain intake by mice, to estimate likely damage to a sunflower crop in New South Wales between the time of survey and harvest. The method assumed no hoarding of food occurred. The biomass of pastures in parts of South Dakota was not significantly different between plots with and without prairie dogs (Collins *et al.*, 1984). There were significant differences in biomass of two species of plants, however. The statistical test used was not specified.

Buckle *et al.* (1984) reported results of a study by Rennison that the frequency distribution of rodent damage to rice fields in a part of Malaysia was log-normal. It was suggested that analysis of percentage damage should use data transformed to logarithms, after adding one, to normalise the data that included some zero values. The effects of rodents in four different rice crops in a part of Malaysia were estimated by analysis of variance by Buckle *et al.* (1984). The analysis showed significant differences in damage between crops, and significant differences between times in crops where rodent control did not occur. Murua & Rodriguez (1989) reported two measures of rodent damage to pines in central Chile; the percentage of pines with damage and the instantaneous rate of damage. The rate was defined as the logarithm of the ratio of successive measures of damage, that is, $\log(D_t/D_{t-1})$, where D_{t-1} is the initial damage and D_t is the damage at the next time of measurement one month later. Lefebvre *et al.* (1989) also applied analysis of variance to examine differences between times in rodent damage to sugarcane in southern Florida.

The effects of farms, times and position in a crop on mouse damage to soybean crops were analysed by split-plot analysis of variance by Singleton *et al.* (1991a). Because of differing initial numbers of damaged pods and a significant positive correlation between damage and pods per plant, the initial number of pods per plant was also used as a covariate in an analysis of covariance. The only significant effect at the transect level was a difference in damage between times of sampling. There were not significant differences between farms.

The analyses used to study rodent damage have not been adequately compared. The comparative biases and cost-effectiveness of methods of damage estimation and of analysis need to be assessed.

The statistical analysis of rodent damage is discussed in section 3.17, economic analysis in section 4.6 and modelling in sections 5.7 and 6.9.

2.7 Bird strikes on aircraft

2.7.1 *Problem definition and species*

Birds and aircraft share airspace but do not always mix. Occasionally, birds and aircraft collide, usually killing the birds and denting the aircraft. Very occasionally the aircraft suffers more serious damage or is destroyed, and passengers are killed (Burger, 1985). One of the earliest airstrikes was described by Solman (1973). A fatal crash occurred in 1912 in northern America after an aircraft struck a bird. The crash was only nine years after the first powered controlled flight by an aircraft.

The problem occurs around the world. Brough & Bridgman (1980) reported that in the United Kingdom in 1972 there were 309 bird strikes with civilian and military aircraft. Human fatalities have occurred after strikes with lapwings (*Vanellus vanellus*) in Britain (Milsom, Holditch & Rochard, 1985) and with starlings (*Sturnus vulgaris*) in Boston, Massachusetts (Burger, 1985). In the latter accident 62 people died. Solman (1978) presents interesting reading of case histories of various strikes from around the world.

Most bird strikes occur at low altitude, during aircraft take-off or landing. Hence, most attention to estimating impacts, and control, is centred around airports. There are many species involved in bird strikes, such as gulls (*Larus* spp.), other waterbirds, waders, diurnal and nocturnal birds of prey and small passerines. Birds weighing less than 900 g caused 94% of strikes in Australia and Papua New Guinea from 1963 to 1971 (van Tets *et al.*, 1977).

2.7.2 *Methods of study*

Three methods have been used to study bird strikes on aircraft.

Pilot records (method 1) were used by Burger (1985) who studied bird strikes at J. F. Kennedy airport in New York. The records were submitted when the aircrew were aware a strike had occurred. Burger (1985) noted that these records estimated only 20% of the real strike rate as aircrews were very busy during take-off and landing and so may not have noticed individual birds striking a wing or other part of the aircraft. Van Tets (1969) used similar pilot records in a study at Sydney airport.

Daily counts of bird carcasses on runways (method 2) were recorded by Burger (1985). Carcasses were also searched for after pilots had submitted a strike report. The aim of the counts was to identify the species struck, and to determine sources of variation in strike rates. Examination of bird carcasses was also used by van Tets (1969) and van Tets *et al.* (1977). A third method of studying bird strikes is computer simulation, which is described in section 5.8.

2.7.3 *Analysis*

Van Tets *et al.* (1977) reported that the damage to aircraft is proportional to the weight of the bird, and in theory to the square of the aircraft speed and the cube of the number of birds present. They noted that the probability of severe damage was higher with increasing altitude, as the aircraft speed was then greater, even though the probability of strike decreased with increasing altitude.

The relationship between bird abundance and bird strikes reported by van Tets (1969) and van Tets *et al.* (1977) is worth closer examination. In the original study (van Tets, 1969), the relationship between the cube root of mean abundance and bird strike was reported to be positive and statistically significant ($r=0.95$, df$=3$, $P<0.05$) for five bird species. My analysis of the data shows the correlation between mean abundance and bird strikes was higher and slightly more significant ($r=0.999$, df$=3$, $P<0.01$), and the correlation between the square root of mean abundance and bird strikes was also high and significant ($r=0.989$, df$=3$, $P<0.01$). Hence, there was nothing special about the cube root relationship, though it had some theoretical appeal.

Van Tets (1969) also reported bird strikes for three other bird species during the same study but did not apparently include them in the analysis. My analysis shows that when data for all eight species are included there is a significant correlation ($r=0.998$, df$=6$, $P<0.01$) between mean abundance and the number of bird strikes and the square root of mean abundance and bird strikes ($r=0.979$, df$=6$, $P<0.01$).

Burger (1985) reported that the analysis of bird strike records was hampered by poor identification of bird species. However, detailed identification and comparison with independent data on abundance showed that herring (*Larus argentatus*) and great black-backed gulls (*L. marinus*) were struck significantly less than expected, and ring-billed (*L. delawarensis*) and laughing gulls (*L. atricilla*) were struck significantly more than expected as assessed by Chi-square analysis. Significantly more young than old gulls were struck. The strike rate estimated from carcasses was always higher than that estimated from pilot reports.

Pilot reports of bird strikes at J. F. Kennedy airport varied seasonally, being highest in May and November, though carcass counts showed no clear seasonal trends. Diurnal variation in strikes of gulls and non-gulls was not related to the number of aircraft movements. Strikes were most commonly reported in early morning but aircraft movements peaked in late afternoon or early evening. Significantly more birds were struck, as assessed by Chi-square analysis, by wide-bodied aircraft such as the Boeing 747 and DC-10, than narrow-bodied aircraft. Burger (1985) found that the number of gulls was not significantly correlated with the number of gull strikes ($r_s=0.18$, df$=14$, $P>0.05$) or the number of gull carcasses ($r_s=0.04$, df$=14$, $P>0.05$). The correlation analysis was a Spearman rank correlation. Different results between the studies may be related to the different evasive behaviours of birds when threatened with a large aircraft or differing aircraft speed and shapes.

The statistical analysis of control of bird strikes on aircraft is discussed in section 3.18, economic analysis in section 4.7 and modelling in sections 5.8 and 6.10.

2.8 Bird damage to crops

2.8.1 *Problem definition and species*

Birds cause damage to crops in many parts of the world; such as corn in northern America, wheat in New Zealand, orchards in Australia and grain crops in Africa. There is a large literature on bird damage and only a small sample is described here. A comprehensive description of crop damage by red-billed quelea (*Quelea quelea*) in Africa was given by Bruggers & Elliott (1989). The review described the ecology of the species, types of damage, assessment and control. In northern America blackbird damage to corn has been studied intensively.

2.8.2 *Methods of study*

Otis (1989) described the two components of bird damage estimation: sampling, which determines sample size, and damage assessment, which is what is done once the sampling points are reached. Each was discussed in detail for crop damage by red-billed quelea in Africa, though the principles apply more widely. The scale of the damage estimation could vary from a field to a survey of a large geographical area.

Bird damage to crops has been studied by four main methods. A questionnaire survey (method 1) was sent by Dawson & Bull (1970) to fruit growers in New Zealand. The survey aimed to determine what types of fruit crop were damaged by what species of birds. The growers were surveyed systematically and asked to rank the bird damage on a scale of no damage, to slight, moderate or severe damage. Wakeley & Mitchell (1981) surveyed, by questionnaire, 1056 corn growers in Pennsylvania and found 45% reported losses to birds.

Direct sampling (method 2) of wheat fields in a part of New Zealand was used by Dawson (1970). Points were selected randomly within the crop and five ears of wheat nearest to the sample point were picked and examined for damage by house sparrows (*Passer domesticus*). Murton & Jones (1973) sampled Brassicae crops in England by randomly selecting 100 plots diagonally across each field. Damage by wood-pigeons (*Columba palumbus*) was scored from zero (no damage) to three (leaves stripped). Damage by skylarks (*Alauda arvensis*) could be distinguished from that by wood-pigeons. Murton & Jones (1973) also sampled the stomach contents of the pigeons to confirm that they were eating the damaged plants (method 3).

Sampling procedures needed for accurate and reliable estimation of blackbird damage to corn fields were discussed by Granett *et al.* (1974). The number of corn ears to be measured was least when damage was high, plant density was low and the number of ears per sample was low. Stickley, Otis &

Palmer (1979) described a three-stage cluster sampling scheme for estimating damage, including that by birds, to corn. The first stage was a county, the second stage a plot of corn within a county, and third stage was a set of corn ears within the plot. Counties were selected with replacement, and with probability proportional to the area of corn harvested in the counties. As several plots were sampled in each county, the plots were clustered. The survey sampling was refined as variability changed, indicating that more counties and fewer plots in each county had to be sampled. Cluster sampling actually required more samples to be measured than simple random sampling, but the greater practicality of cluster sampling may have outweighed the statistical limitation. This result is not unique to studies of bird damage to corn. Sudman *et al.* (1988) reported a similar increase in sampling variance with cluster sampling compared with simple random sampling in social surveys.

The relationship between insect abundance in sweet corn and damage by red-winged blackbirds was examined by Straub (1989). Two corn crops were planted experimentally in New York and within each crop, ears of corn were randomly sampled for infestation by European corn borer. Separate samples of corn ears were examined over different times to estimate the accumulated percentage of ears damaged by the blackbirds. Whether the samples were selected randomly was not specified. Finally, a random sample of ears damaged and undamaged by blackbirds was collected and examined for corn borers. One corn crop was treated with insecticide and the other crop was an experimental control, so there was no replication of treatments.

The behaviour of red-winged blackbirds on two cultivars of sweet corn was reported by Bernhardt & Seamans (1990). Captive birds were offered one corn variety then both followed by offering the second variety alone. That experimental arrangement was used because of practical limitations on the number of corn heads available. An alternative experimental procedure may have randomised the sequence of offerings.

Long (1985) used direct sampling of trees to estimate the mean percentage of fruit damaged by parrots in south-western Western Australia. Initially, every tree in the orchards was examined, but in later work every fifth tree was examined so sampling was systematic.

Johnson *et al.* (1989) randomly selected aerial photographs of their study area in Texas, then selected the grapefruit grove closest to the centre of the photograph. Within each grove the first tree was randomly selected and subsequent trees sampled systematically. Sample sizes and trees per grove were determined by a preliminary survey to obtain precise estimates of damage by grackles (*Quiscalus mexicanus*), though the actual level of precision required was not stated.

Simulation modelling of bird damage to crops (method 4) is described in section 5.9.

The different methods have potential biases and advantages. Otis (1989) considered that all damage assessment methods were to some extent subjective, but that bias could be reduced by training observers. Dawson & Bull (1970) reported that their questionnaire survey had a low return rate; 10% of pip and stone fruit growers and 22% of berry growers replied. There was a tendency for growers with the worst damage to reply and for growers with least damage not to reply. This was established by resurveying a random sample of the growers. The survey also had problems with mis-identification of birds by growers. A potential problem not discussed by the authors is that a moderate damage problem to one grower may be a severe problem to another. Weatherhead, Tinker & Greenwood (1982) considered that direct sampling had two main limitations – extensive field sampling was necessary, and time was limited to do that sampling. Otis (1989) considered that when sample size was fixed in direct sampling it was more efficient to sample many crops rather than sample fewer crops more intensively.

2.8.3 *Analysis*

The extent of bird damage to crops has been estimated as the absolute amount (kg) (Weatherhead *et al.*, 1982), the percentage of crop lost, the economic cost (Dolbeer, 1981) and the incidence of damage (Otis, 1989). Ideally, damage would be estimated in units of lost production, for example kilograms, and that converted into monetary units.

Dawson (1970) reported that the spatial frequency distribution of damage by house sparrows to a wheat field in New Zealand was positively skewed, so was transformed to a normal distribution by a logarithmic transformation. It was also reported that compensatory growth was not significant as assessed by a paired t test of grain weight from undamaged and damaged ears of wheat. Further study showed that compensatory growth occurred if the grains were removed during the early 'milk' stage of development but not if they were removed later. The efficiency of different sampling efforts was examined. The variance of a damage estimate differed curvilinearly with the number of ears sampled. The variance decreased when the number of ears sampled was small but as the number of ears increased then the variance also increased.

The spatial variation in the level of blackbird damage is highly skewed being of a negative exponential form. Dolbeer (1981) and Wakeley & Mitchell (1981) reported that most corn crops had no or little damage and few crops had high levels of damage. The pattern of variation was very similar to that described for predation of sheep (section 2.4). Dolbeer (1981) suggested that the spatial variation was associated with differing distances from blackbird roosts. Damage declined exponentially as distance increased. The relationship was suggested as a method for predicting damage but was not tested independently.

Murton & Jones (1973) analysed for differences between years in wood-pigeon damage to Brassicae crops in parts of England. They used a Mann–Whitney *U*-test on the damage data, which were ranked scores of the damage. Significant differences in pigeon damage between years were reported.

Red-winged blackbirds (*Agelaius phoeniceus*) feed by shredding the enclosing corn sheath (husk) and pecking at the developing corn kernels. The blackbirds are territorial during the breeding season and assemble in large roosts after breeding. It is during such roosting that corn damage is most severe (Wiens & Dyer, 1975). Dyer (1975) reported that the corn ears eaten by blackbirds were generally longer than those not eaten. Analysis of covariance, using ear length as the covariate, showed that the weights of damaged ears were significantly greater than those of undamaged ears. The longer ears tended to be insufficiently covered by the enclosing sheath and so were selected for feeding. Such feeding may also act as a stimulant for plant compensatory growth, sometimes resulting in an increased corn yield. Dyer (1975) provided data from Ontario and Ohio that supported the hypothesis of plant compensatory effects. Woronecki, Stehn & Dolbeer (1980), however, concluded that the effect on yield of plant compensation was generally insignificant, particularly if the secondary effect of bird attack such as fungal infection and insect damage were considered. Woronecki & Dolbeer (1980) further examined insect damage and bird abundance. They concluded that although blackbirds feed on insects in corn crops, studies had not demonstrated the economic benefits of blackbirds feeding on insect pests in corn.

The effect of sweet corn variety on behaviour of red-winged blackbirds was analysed by Chi-square analysis (Bernhardt & Seamans, 1990). Significant differences in the frequency of behaviour to access the corn kernels were observed. Differences in kernel losses were significant as assessed by paired *t* test in both one-choice and two-choice tests. Relations between corn borer abundance and damage to corn by red-winged blackbirds were estimated by Straub (1989), though no results were statistically analysed.

Damage to grapefruit in a part of Texas was negatively correlated ($r = -0.97$, df $= 3$, $P < 0.01$) with distance from the edge of a grove (Johnson *et al.*, 1989). Regression analysis showed that grapefruit damage was significantly and negatively related to distance from sugarcane (closer fields had more damage), positively related to distance to the nearest grove (the further the grove the higher the damage), and negatively related to the size of a grove (smaller fields had more damage). These regressions were calculated independently. An extra analysis would be multiple regression to determine the combined contribution of distances and size to damage.

Long (1985) reported damage by birds to fruit trees in six orchards in Western Australia but did not statistically analyse the results. That is done here for a subset of the results. The experimental design is factorial with three factors: orchards, years and months. The factors, orchards and years, were each at two levels and the factor months was at five levels, giving a total of 20 ($=2\times2\times5$) treatment combinations. Orchards were selected on the basis of having reported previous damage so were not a random sample of orchards. The selection of months was presumably not random as fruit does not occur on trees year round. The basis for selection of years was not described.

Data from Table 2 in Long (1985) are presented here in Table 2.5. The data are from two orchards (B,C) at Balingup in 1974 and 1975. The data can be analysed by three-way fixed-factor analysis of variance (Snedecor & Cochran, 1967). The variable analysed, after arcsine transformation, is the percentage of fruit damaged.

The average damage on orchard B (20.0%) was significantly different from that (10.0%) on orchard C ($F=9.44$; df$=1,4$; $P<0.05$). There were significant differences between months ($F=6.79$; df$=4,4$; $P<0.05$) in the level of damage, being highest in February and lowest in May. There were no significant effects of years or the two-way interactions of any of the treatments. The three-way interaction (year $\times$ month $\times$ orchard) could not be tested because of lack of replication. No estimates of the occurrence or effects on yield of compensatory growth by fruit trees were made in this study. Interpretation of the above analysis could be limited by lack of independence between treatments. For example, damage in February may be dependent on the level of damage in January.

Table 2.5. *Damage to fruit by parrots in south-western Western Australia*

		Damage (% of fruit)				
Year	Orchard	Jan.	Feb.	Mar.	Apr.	May
1974	B	10.9	71.1	8.0	7.1	2.9
	C	45.3	43.5	11.0	0.2	0.1
1975	B	3.4	43.9	31.2	21.2	0.0
	C	0.0	0.0	0.0	0.0	0.0

Data: From Long (1985).

The methods and analyses described show that the variability in bird damage has been more often studied than other topics, with the exception of livestock predation. Many studies have used standard statistical analysis but some studies persist where no analysis has occurred.

The statistical analysis of control of bird damage is discussed in section 3.19, economic analysis in section 4.8 and modelling in sections 5.9 and 6.11.

2.9 Rabbit damage

2.9.1 *Problem definition and species*

Rabbits (*Oryctolagus cuniculus*) in southern Australia have been accused of reducing agricultural production through grazing competition with livestock, and changing the composition of native vegetation. Sumption & Flowerdew (1985) described many studies of the ecological effects of rabbits in Britain. Sheail (1991) considered that an earlier lack of data on damage by rabbits had inhibited management of the species.

2.9.2 *Methods of study*

Three methods of study of the effects of rabbits on vegetation have been used: enclosures, exclosures and analysis of plant age structure.

The first method was used by Myers & Poole (1963) who placed rabbits in enclosures to determine the effects on sown pastures in south-eastern Australia. Exclosures were placed inside each enclosure to separate grazing and seasonal effects. Different enclosures had different initial numbers of rabbits. Lange & Graham (1983) used enclosures to examine the survival of four species of *Acacia* in South Australia. Seedlings were transplanted into four fenced enclosures into which rabbits were later placed. Equal numbers of seedlings and equal numbers of rabbits were placed in each enclosure. There was no experimental control (seedlings with no rabbits). The survival of the seedlings was noted every 12 hours for up to 144 hours. Croft (1990) used enclosures in a randomised block design with different rabbit densities, including zero, as treatments. The effects of rabbits on pasture and sheep production were estimated. There were four levels of rabbit abundance and each was replicated.

The second method was used by Cooke (1987) to study the effect of rabbit exclusion on survival of seedlings of trees, *Allocasuarina* and *Melaleuca*, in Coorong National Park in South Australia. One group of seedlings was planted within netting tree guards, and another group planted within guards that allowed rabbits to enter and leave but prevented entry by larger mammalian herbivores such as kangaroos and wombats. Damage to the seedlings of each species was scored on a scale of 0 (no damage) to 4 (most damage) at each time of inspection. The duration of each experiment was about seven to ten months. The two plant species were planted in separate exclosures and the differences examined at differing times of year. Foran, Low & Strong (1985) used exclosures to study the effects of rabbits on vegetation in central Australia.

The third method involves determining the age structure of a plant population to determine if the net effects of recruitment and survival show any periods of low or very low net survival. The method requires estimation of a relationship between size and age, but does not require long-term experiments using enclosures or exclosures. The method was used by Crisp & Lange (1976) to study survival under grazing of a shrub, *Acacia birkittii*, in Australia.

2.9.3 *Analysis*

Myers & Poole (1963) did no statistical analysis of their data. My analysis of spring biomass data in their Tables 6 and 7 is to test for effects of three factors: rabbit grazing (with rabbits and with no rabbits), years (1957 and 1958) and plant taxa (five groups). The factors of grazing and plant taxa were assumed to be fixed and that of years was assumed to be random. A three-way analysis of variance, after transformation to common logarithms to improve variance homogeneity, showed a significant effect of plant taxa ($F=10.53$, df$=4,4$, $P<0.05$) but no significant effects of rabbit grazing ($F=4.90$, df$=1,1$, $P>0.05$), years ($F=2.14$, df$=1,4$, $P>0.05$) or any of the two factor interactions. The analysis used the three-way interaction (rabbits $\times$ plants $\times$ years) as the residual variance as there was no replication of the three-factor combination. Replication occurred in the original experiment but means, not the original data, were presented in Tables 6 and 7.

Myers & Poole (1963) concluded that rabbits decreased pasture biomass but this analysis does not support the conclusion. Similarly, they concluded that rabbits were an additional form of grazing to livestock, even though they did not test the hypothesis or have sites with rabbits and livestock. Rabbit numbers varied in each enclosure so a possible improvement in the above analysis would be to use rabbit numbers as a covariate in an analysis of covariance.

Lange & Graham (1983) simply calculated the percentage of *Acacia* seedlings still alive at regular intervals, and plotted the trends. There was no statistical analysis of the data. The trends, of a rapid loss of seedlings, were very clear, however, in three enclosures and there was no effect in the fourth, apparently associated with abundant alternative grass. Cooke (1987) similarly simply tabulated the frequency of occurrence of damage scores with increasing time since planting. In one experiment with *Allocasuarina* all the plants were severely damaged within two days. It appeared from interpretation of the age structure that *Allocasuarina* began to regenerate early in the 1950s during the period of low rabbit abundance following the initial epizootic of myxomatosis. These studies suggest a greater need for statistical analysis of data on rabbit damage. In contrast, Croft (1990) tested the effects of rabbits on pasture and sheep production by analysis of variance and showed that as rabbit density increased, the species composition of pasture changed significantly and sheep production declined significantly. The analyses described here have often been nothing more than eye-balling exercises, so the study by Croft (1990) is a welcome change.

The statistical analysis of rabbit damage control is discussed in section 3.20, economic analysis in section 4.9 and modelling in sections 5.10 and 6.12.

2.10 Conclusion

Estimating the damage of vertebrate pests has used well-established principles in experimental design and sampling through to careful observation of the behaviour of the animals concerned. The case studies described here show that some of these principles have been used, but there is scope for greater use of random sampling and genuine replication. Similarly, studies rarely describe the frequency distribution of damage or attempt to identify sources of

Table 2.6. *Checklist of steps in estimating and analysing the damage by vertebrate pests*

1. Investigate whether species is a cause of concern		
	(i) Indirect assessment	– questionnaire
		– interviews
	(ii) Direct assessment	– behaviour
		– food habits
2. Estimate type and level of damage		
	(i) Type	– direct (consumption)
		– indirect (contamination)
	(ii) Mean level	– physical units (kg/ha, number)
	(iii) Level of variation	– variance, frequency distribution
	(iv) Experimental design and sampling	
		– schemes
		– randomisation and replication
	(v) Significance of results of statistical analysis	
	(vi) Comparison of observed to expected (modelling) results	
3. Extrapolations		– laboratory to field
		– small plots to large areas
4. Identify spatial and temporal sources of variation in levels of damage		
	(i) Pest abundance	
	(ii) Environmental factors	

variation in the frequency distribution. The detailed study of rabbit grazing on wheat by Crawley & Weiner (1991) is a rare exception. Surprisingly, many studies, including recent ones, persist in giving results but not doing any statistical analysis. My analysis of several such data sets shows that the conclusions of the original studies and the analyses are not always the same.

Over the range of topics in this chapter, there are often many methods for estimating damage and for statistical analysis of such estimates. There have been some comparative evaluations of the methods and analyses. One example of comparative study was by Schaefer *et al.* (1981) in an investigation of methods of estimating livestock predation.

A suggested sequence of tasks involved in estimating damage by vertebrate pests is summarised in Table 2.6. The list should be a framework for answering the questions at the start of this chapter, such as those about the damage by goats on the semi-arid shrubland and other vertebrate pest problems. A more detailed checklist of steps common to the analysis of damage and control is given in Table 3.7 and should be used in conjunction with Table 2.6.

When pest damage has been clearly identified then planning of the analysis of damage control can be more advanced. The statistical analysis of the response of damage and pests to control is the subject of the next chapter.

3

Statistical analysis of response to control

Vertebrate pest control is replete with stories of the wily trapper, pitting his or her experience and knowledge against the sly predator. But what are the effects of trapping on predator abundance or the damage by the predators? This chapter is concerned with methods and analyses for determining both. As in Chapter 2, many of the principles come from experimental design, but the applications here are different. The chapter briefly examines the effects of control on the frequency distribution of damage or pest abundance, then reviews statistical analyses used. A variety of control methods, such as poisoning, trapping and fencing, then control topics, such as predation and infectious diseases, are reviewed. The control topics correspond to those described in Chapter 2 with two extra topics.

Conway (1981) classified five alternative techniques for pest control: (i) pesticides, (ii) biological control, (iii) cultural control by use of practices to alter habitats, (iv) resistance such as breeding of crop varieties that are more resistant to pest attack, and (v) sterile mating control to reduce fertility possibly through genetic methods.

The relevance of the techniques to pest species varying from *r*-selected pests to intermediate to *K*-selected pests was described by Conway (1981). Species with high birth rates, more variable abundance, amongst a variety of other population characteristics, are *r*-selected species, and those with low birth rates and less variable abundance are *K*-selected species. Pesticides, cultural control and resistance were considered useful for all such pests but biological control was suggested for use only on intermediate pests and sterile mating control for *K*-selected pests. The framework may be useful for selecting alternative actions in the general planning process. A difficulty in use of the framework is in deciding whether a particular pest is an *r*-selected, *K*-selected or intermediate pest. Caughley (1976) considered, on the basis of modelling herbivore dynamics, that the *r* and *K* classification was simplistic.

3.1 Effect of control on damage and pests

The analysis of response to control is concerned primarily with determining the direction and magnitude of changes of damage or pest abundance after control. The average response or the frequency distribution of responses is usually studied. Most control, lethal or non-lethal, reduces pest abundance or causes a change in pest distribution. The theory of pest damage described in section 5.4 suggests that a reduction in pest abundance will reduce pest damage, though the response may be linear or non-linear.

Control work may influence parameters of the pest population such as sex ratio and age structure. Green & Coleman (1984) described the demographic characterisics of a population of brushtail possums (*Trichosurus vulpecula*) in New Zealand that had been subject to intensive trapping. The population was compared with a nearby non-controlled population. The sex ratio in the trapped area, as assessed by Chi-square analysis, was significantly biased towards males, because of a large number of males in the four-year-old age class. The age structures of the trapped and non-trapped populations, as assessed by Chi-square analyses, were also different, as the non-trapped population contained more older animals. In the area previously trapped, three-year-old males were significantly heavier than in the non-trapped area, as assessed by analysis of variance. There were however no significant differences, as assessed by Chi-square analysis, in the proportions of breeding females in the two populations.

Estimating the effect of control on the average level of damage or pest abundance is straightforward, but is no picnic, and will be described in more detail in section 3.4. Estimating the effect of control on the frequency distribution of damage or pest abundance is not so simple nor as often investigated. Possible frequency distributions of damage were shown in Fig. 2.1. The emphasis here is on what happens to the frequency distribution when control occurs. If the spatial frequency distribution of damage is highly skewed and control occurs equally in all areas, then the expected response is to lower the level of damage at all sites (Fig. 3.1a). When control occurs only in areas of greatest damage then the response will be unequal and will be directional (Fig. 3.1b). Similarly if control occurs only in areas where access is possible then the response may be disruptive (Fig. 3.1c), creating a bimodal frequency distribution. These responses are analogous to directional and disruptive selection in genetics (Krebs, 1985). The frequency distribution of damage may be normally distributed, (Fig. 3.1d) and in response to control occurring at all sites, the plot of the distribution may shift to the left.

If damage by a pest can change the mean, variability, skewness (departure from a bell-shaped normal distribution) and kurtosis (a measure of pointedness of a distribution such as having a higher top or longer tails than a normal distribution) of a variable, as reported for rabbit grazing on winter wheat (Crawley & Weiner, 1991), then pest control may do the same. This is an area for future research.

3.2 Effect of pests on control

Control work can influence population parameters of the pests and the reverse is also true – characteristics of the pest population influence control

Fig. 3.1. Possible responses of the frequency distribution of damage to pest control occurring at (*a*) all sites where damage occurs, (*b*) only at sites of highest damage, (*c*) only at sites of access and intermediate levels of damage, (*d*) at all sites when the frequency distribution of damage is normally distributed. The solid line is the frequency distribution of damage in the absence of control, and the dashed line the frequency distribution after control.

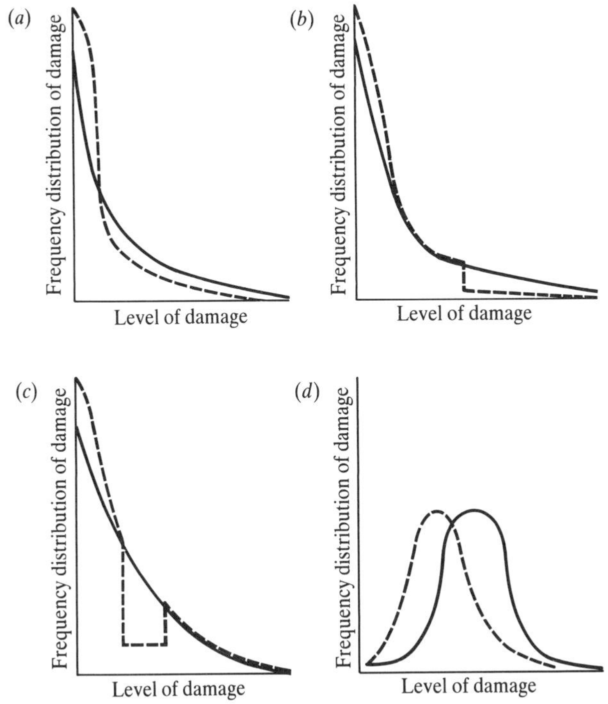

effectiveness. A classic example in rodent control work is the pesticide-resistance status of the population (Greaves, 1985). This topic is discussed further in sections 3.3 and 3.5. Another example is the development of resistance to an introduced pathogen after its use as a biological control agent, and this is discussed in more detail in sections 3.3 and 5.6.

A related issue is the possible co-evolution of species with a naturally occurring toxin and their higher tolerance of that compound than that of introduced pest species. Mead *et al.* (1985) and Twigg & King (1991) discussed this for fluoroacetate, which is produced naturally by many species of legumes in northern and Western Australia. Some native mammals in that state are more tolerant of the compound than the same species in parts of Australia where the fluoroacetate-producing plants do not occur. Introduced pests, such as foxes, which occur throughout southern Australia are not as tolerant of fluoroacetate, a condition that has been used to advantage in south-western Australia where sodium monofluoroacetate (compound 1080) has been used to control foxes (Kinnear, Onus & Bromilow, 1988).

3.3 Spatial and temporal aspects

3.3.1 *Spatial*

The spatial spread of pests may be limited by barriers such as fencing or areas of intensive control. The analysis of the effectiveness of such barriers is described in section 3.7. The use of barriers to limit spread of an undesirable pathogen has also been described (Kallen, Arcuri & Murray, 1985). The estimation by modelling of the width of barriers is discussed in section 6.8.3.

Where control occurs, or where it should occur, can be documented to assist in planning control. A suitable approach to documentation is the use of a geographic information system (GIS). This can combine the data on location, timing and method of damage or pest control. An example of a suitable GIS was described by Kessell, Good & Hopkins (1984), though the GIS was developed for management of fire in natural areas and did not include vertebrate pest data. Coulson (1992) described a GIS for integrated pest management, though again it was not specifically concerned with vertebrate pest control.

3.3.2 *Temporal*

Resistance to the effects of pesticides used for vertebrate pest control has developed in many parts of the world (Greaves, 1985). Most study of the topic has concerned resistance to an anticoagulant pesticide, warfarin. Greaves (1985) described resistance as the failure of a pesticide to control an infestation when used in a normally efficient manner. Resistance is believed to occur naturally in some species, such as *Acomys cahirinus*, and has been selected for and hence increased in others (*Rattus norvegicus*) (Greaves, 1985). The time for resistance to develop appears to be related more to the generation length of a pest than to the frequency of resistant genes and the strength of selection (Dobson & May, 1986). That is, resistance develops faster in species with shorter generation intervals.

Since the introduction of myxoma virus to Australia in 1951 there has been a decrease in the mortality of rabbits as a result of repeated epizootics (Fenner & Ratcliffe, 1965). This was suggested to be because of higher immunity of the rabbit population, though it was apparently also related to the evolution of strains of differing virulence (May & Anderson, 1983; Dwyer, Levin & Buttel, 1990). A crude analysis of the coevolution of changes in the virulence of myxoma virus and the resistance of rabbits in Australia was outlined by Anderson & May (1982a). They reported that high virulence was associated with a low recovery rate from infection and that the highest levels of secondary infection occurred at intermediate levels of virulence of the disease myxomatosis. Higher and lower levels of myxoma virulence had lower levels of secondary infection (estimated by the basic reproductive rate; see section 5.6).

Ross & Tittensor (1986) reported that the rate of spread and the proportions of rabbits infected with and dying from myxomatosis in Britain had declined since the disease was first recorded in British rabbits in 1953. These differences in spread rates were associated with a high initial density of susceptible rabbits. The infection rate and mortality were then lower because of the later higher incidence of immune rabbits.

There is an emphasis in this chapter on analyses for estimating the effects of control at a particular time. However, estimating the effect over time is often required and can be done by calculating the rate of increase of pest abundance. Estimating the rate of increase of a pest population is described in the next section.

3.4 Evaluation of control

The evaluation of control involves the same principles of design, sampling and analysis as used in the evaluation of the damage of vertebrate pests (Chapter 2) but will not be elaborated again here. Rather, in this section, the application of the principles will be further outlined. Evaluations of vertebrate pest control are generally examples of controlled or manipulative experiments, with or without replication, though preferably with. Hence, the discussions of Eberhardt & Thomas (1991) and Skalski & Robson (1992) on design and replication are again very relevant.

A comparison of the effects of control on damage or pest abundance may be between sites and populations at the same time (simultaneous) or at subsequent times (sequential) (Fig. 3.2). Sequential and simultaneous comparisons should have fewer possible confounding effects of time and seasonal conditions. A lack of an experimental control is almost an example of what Eberhardt & Thomas (1991) call intervention analysis – that applied to events when a distinct perturbation occurs but when the causes of the perturbation are not fully controlled by the observer.

Examples of the analyses used in the evaluation of vertebrate pest control are listed in Table 3.1. Each analysis could also be used in the evaluation of damage by vertebrate pests. The analyses in Table 3.1 are tests of difference, except for the regression and multiple regression analyses, which are analyses of response.

A detailed comparison of the biases and efficiencies of the various analyses has not been reported. In general, analysis of variance (ANOVA) is more powerful and robust than Chi-square analysis or a Student's *t* test (Snedecor & Cochran, 1967). ANOVA also allows simultaneous testing of more than two treatments, compared with only two for the *t* test. Huson (1980) compared three methods (6, 7, 8 in Table 3.1) by computer simulation and concluded that method 6, the post-treatment census taken as a percentage of the pre-treatment census, was more accurate. The use of the pre-treatment census data as a covariate (analysis 8) was slightly less accurate. That result may not be universal. Woollons &

Fig. 3.2. The difference between sequential and simultaneous experimental controls in evaluation of vertebrate pest control. Cells from which data are collected are shown by solid lines; those from which no data are collected are shown by dashed lines. (*a*) A sequential experimental control occurs where data are collected before and after pest control but only at sites where pest control is about to occur or has occurred. (*b*) A simultaneous experimental control occurs when data are collected after pest control has occurred, at sites where pest control did and did not occur. (*c*) Sequential and simultaneous experimental controls occur when data are collected before and after pest control at sites where pest control did and did not occur.

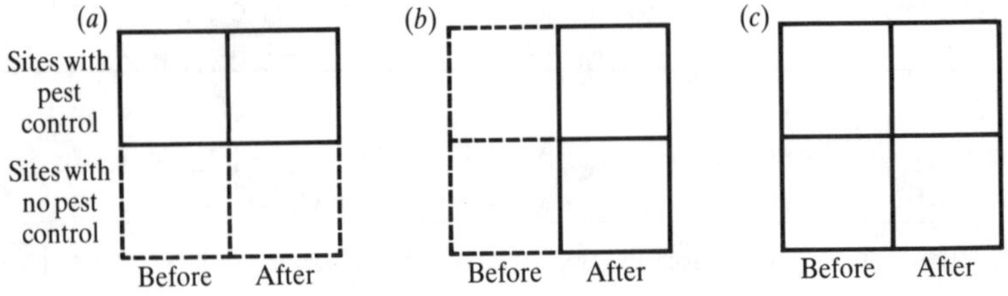

Table 3.1. *Examples of statistical analyses of the response to control of vertebrate pest damage or abundance. The type of experimental control is also listed. See Fig. 3.2 for explanation*

No.	Analysis	Experimental control	Damage or abundance	Source
1.	Chi-square	Sequential	Damage (predation)	Bjorge & Gunson (1985)
			Abundance	O'Brien (1988); O'Brien, Lukins & Beck (1988)
		Simultaneous	Abundance	Sinclair & Bird (1984)
			Damage (trees) and abundance	Sullivan *et al.* (1988b)
		Sequential and simultaneous	Abundance	Twigg *et al.* (1991)
2.	Independent *t* test	Sequential	Abundance	Hone (1983)
3.	Mann–Whitney *U*-test	Simultaneous	Damage (crops)	Murton & Jones (1973)
		Sequential	Abundance	Lazarus & Rowe (1982)
4.	Analysis of variance (post-control)	Simultaneous	Damage (trees)	Murua & Rodriguez (1989)
			Abundance	Cooke (1981); Cooke & Hunt (1987)
5.	Analysis of variance (pre- and post-control)	Sequential	Abundance	Hone & Pedersen (1980)
		Sequential and simultaneous	Damage (soybeans)	Singleton *et al.* (1991a)
6.	Analysis of variance (post-/pre-control %)	Sequential	Abundance	Huson (1980); Hone & Atkinson (1983); Buckle *et al.* (1984)
7.	Analysis of variance (pre- minus post-control)	Sequential	Abundance	Huson (1980)
8.	Analysis of covariance (pre-control as covariate)	Sequential	Abundance	Rennison (1977); Huson (1980); Cooke (1981); Foran *et al.* (1985)

Table 3.1. (*cont.*)

No.	Analysis	Experimental control	Damage or abundance	Source
9.	Regression	None	Abundance	O'Brien (1988); Hone (1990)
		Simultaneous	Abundance	Choquenot *et al.* (1990)
10.	Multiple regression	None	Abundance	Richards & Huson (1985)
		Simultaneous	Damage (oilseed rape)	Inglis *et al.* (1989)
11.	Generalised linear models	Sequential and simultaneous	Abundance	Hone & Kleba (1984)
12.	Multivariate analysis	Sequential	Damage (pasture)	Foran (1986)

Whyte (1988) described the use of covariance in analysing forest fertilizer experiments and suggested that multiple covariance may be very useful because of the increased precision of an analysis.

Skalski & Robson (1992) review statistical tests that have been or could be used to estimate the effects of manipulative experiments where wildlife abundance was estimated by mark–recapture methods. The analyses were not specifically used for vertebrate pest research but could be. Many analyses were limited in use by no or little replication of treatments applied in the experiments.

The analyses listed in Table 3.1 have not been thoroughly evaluated for potential problems in analysis. For example, in a study of the effect of poisoning on abundance of feral pigs (Hone, 1983), the effect was analysed by Student's independent t test. The study used a factorial design with one treatment (poisoned) site and one simultaneous experimental control (not poisoned) site. Replicate counts of feral pigs were obtained within each site. Hence, there were no treatment replicates and the repeated counts are an example of Hurlbert's pseudoreplication. The phenomenon is probably as widespread in vertebrate pest research as Hurlbert (1984) reported for ecological research.

If the evaluation of control is concerned only with the sites and populations under study and the research involves replicated experimental controls, then the data may be amenable to analysis of variance with use of fixed effects models. In contrast, if the evaluation is concerned with certain sites or populations as a random sample of a larger population then random effects models of analysis of variance are appropriate. The distinction is more than academic. The robustness of the analyses is different (Eberhardt & Thomas, 1991). The fixed effects models are less sensitive to violations of the assumptions of analysis of variance. The derivation of the expected mean squares for each model is described in Steel & Torrie (1960) and Snedecor & Cochran (1967).

The evaluation of control may need to distinguish between control inputs applied at a single time, or alternatively at several times over a response period. In the latter case control inputs may be constant or variable. Similarly, responses to control may be classified as occurring instantaneously or delayed, or over a short time period or over a long time period. Evaluation of the temporal effect of control on pest abundance often involves estimating the observed rate of increase of the pest population. Caughley (1980) described how the observed instantaneous rate of increase can be estimated, as the regression of abundance (transformed to natural logarithms) and time.

An example of analysis of trends was reported by Hone & Stone (1989), who estimated the rates of increase of two feral pig populations, one in areas of no control and the other in areas where poisoning, with warfarin, occurred. The statistical test was of the slopes of the regressions between the natural logarithms of abundance (y) and time (x). The hypothesis was that the slopes were zero, which occurs when each population is stable. An alternative analysis would be to compare directly the slopes of the regressions using analysis of covariance (Snedecor & Cochran, 1967).

Harris (1986) showed that when estimates of animal abundance are highly variable it was necessary to obtain multiple counts each year to ensure a precise estimate of rate of increase. More surveys give an analysis greater power to detect a given rate of increase and higher rates of increase are easier to detect (Gerrodette, 1987). Eberhardt & Simmons (1992) further discuss the estimation of rate of increase and imply that earlier analyses (Harris, 1986; Gerrodette, 1987) may have underestimated sample sizes. Harris (1986) also examined the reliability of analyses of rates of increase. The log-linear regression analysis, as described above, was robust to violation of assumptions of exponential growth, independence of counts and deviations from a log-normal frequency distribution. However, the precision of estimates of rates of increase were not always robust to violation of assumptions. Similarly, if the bias of a counting method changed over time then estimates of rates of increase would be biased.

The analysis of response to pest control may give ambiguous results. For example, eradication of feral goats from an island may not result in an increase in distribution and abundance of a particular plant species or a community of plants. The recovery may not produce a plant community similar to that which existed prior to introduction of the feral goats. Schofield (1989) reported such an apparent response of some plants to eradication of feral goats from an island in the Galapagos. The lack of response may be because of a lack of seed bank and/or limited dispersal from areas where plants survived. Nugent (1990) suggested that a reduction in abundance of introduced fallow deer (*Dama dama*) in forests in New Zealand may result firstly in increases of less palatable species and highly palatable species would increase only if deer abundance was greatly reduced. Trapping was reported to reduce rat (*Rattus* spp.) abundance in macadamia nut plantations in Hawaii but trapping did not reduce damage (Tobin *et al.*, 1993). The result may have been associated with plant compensation or small sample sizes used to estimate nut yield, as discussed by the authors.

The procedures used to evaluate several methods, such as poisoning or trapping, of vertebrate pest control are now described (sections 3.5–3.14). The procedures and the analyses are not intended to be exhaustive but to illustrate a range of good and bad examples. The procedures are then described for various vertebrate pest topics, such as predation control and control of infectious diseases (sections 3.15–3.22), corresponding to the topics discussed in sections 2.4 to 2.9.

3.5 Poisoning

The evaluation of poisoning can involve one or more methods of study, which are now discussed. Use of the methods may often be in the chronological sequence listed. The methods could be used for both the pest (target) and non-target species.

3.5.1 *Laboratory tests of toxicity*

This usually involves treating pest animals with different doses of a pesticide and recording the mortality at each dose level. The poison may be given orally, applied to the skin in a dermal test, injected (subcutaneous – under the skin, intraperitoneal – into the peritoneal cavity, or intramuscular – into the muscle), or offered in water or food.

The procedure often estimates the relationship between the dose of poison and the response, for example percentage kill, and from the relationship estimates the dose needed to kill 50% of a population (the LD_{50}), with 95% confidence intervals (Fig. 3.3). The LD_{50} is also called the median lethal dose. Hone & Mulligan (1982) tabulated toxicity data for 38 pesticides used in vertebrate pest control.

Data for several dose levels are required with replication of animals at each dose level. A minimum would be three dose levels with two animals at each level. The estimated response from such minimum data would, however, be expected to be very imprecise and nearly useless because of the small sample size. If the experiment is using rare species such sample sizes may be imposed, so a non-lethal response may be more useful. Woods (1974) suggested that the efficiency of dose–response testing could be improved if more animals were given a dose nearer the expected LD_{50} level. Obviously, that requires some prior knowledge of the likely response. The statistical methods of estimating an LD_{50} are not reviewed here. However, two methods have been frequently used: moving averages (Thompson, 1947; Weil, 1952) and probit analysis (Finney, 1971).

Standardised procedures are necessary for laboratory testing of toxicity. McIlroy (1981) examined the effects on susceptibility to sodium monofluoroacetate (compound 1080) of animal age, sex, nutritional state, location, route of administration and the number of animals used at each dose level. Limited numbers of animals restricted the statistical analysis of the results to an independent Student's *t* test for each treatment. Ideally, an analysis should have tested for interactions between treatments. On the basis of the results, and others reported in the literature, McIlroy (1981) planned to use, in future toxicity studies, adult males provided with continual access to food and water, from defined locations, of close to average weight and that do not suffer stress in captivity, to avoid extremes of temperature, use five animals at each dose level, dose all animals by the oral route, use standard poison solutions in a small volume with administration of the dose at the same time of day by the same people. Redosing of animals was to be avoided and the results should be

Fig. 3.3. The relationship between the response to poisoning, such as percentage kill, and the logarithm of the dose of poison. (After Woods, 1974.)

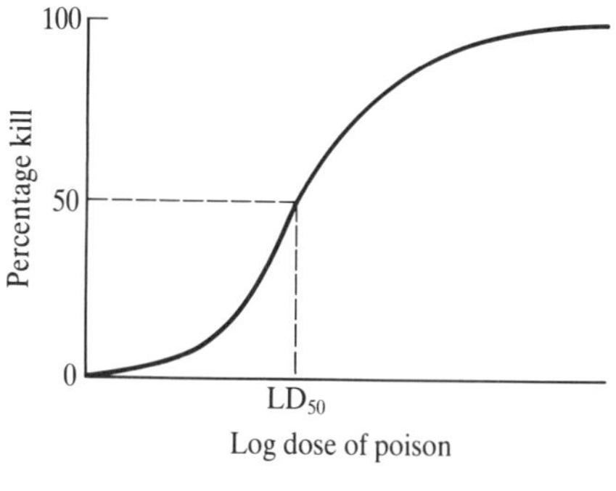

used in a single type of analysis to estimate an LD_{50} and the 95% confidence interval.

The effects of ambient temperature were further studied by Oliver & King (1983), who reported substantial effects of temperature on the toxicity of sodium monofluoroacetate (compound 1080) to mice (*Mus musculus*), guinea pigs (*Cavia porcellus*) and brushtail possums (*Trichosurus vulpecula*). The toxicity of 1080 to mice and guinea pigs was lowest (higher LD_{50}) at intermediate temperatures (24°C). The effects of temperature were not statistically tested, though the lack of overlap in most 95% confidence intervals suggests the conclusions were correct. Eastland & Beasom (1986) reported an effect of temperature on the toxicity of 1080 to raccoons (*Procyon lotor*). Toxicity was higher (lower LD_{50}) at 23–37°C than at 13–23°C. The difference was reported to be significant ($P < 0.05$), though no details were given of which statistical test was used. Raccoons were held outdoors in individual cages for the experiment. Temperatures, an important experimental factor, were recorded, surprisingly, about one kilometre from the study site.

The results of laboratory testing may not be easily applied to field situations. It may be of greater practical value to estimate the LD_{90}, the concentration of poison in a bait that kills 90% of a population, but such statistics are estimated with lower precision (wider confidence intervals) than the LD_{50}. The differing conditions between a laboratory and the field, such as temperature or presence of alternative food, may alter the toxicity of poisons or acceptance of baits (McIlroy, 1986). Sinclair & Bird (1984) showed that some *Sminthopsis crassicaudata*, a marsupial mouse in Australia, ate only small amounts of meat baits containing 1080, compared with unpoisoned meat, though the average difference in intake in a one-choice (poisoned meat) or two-choice (poisoned and non-poisoned) test was not significant in analysis of variance.

3.5.2 *Laboratory tests of bait acceptance and physiological responses*

This involves laboratory assessment of the non-toxic part of poisoning, such as bait acceptance. Kleba, Hone & Robards (1985) studied the acceptance of three grains (wheat, maize and sorghum), dyed and undyed, by penned feral pigs. Undyed grains were offered simultaneously for three days and then dyed grains offered for three days. This sequential arrangement simulated a hypothetical field practice of offering undyed (unpoisoned) bait followed by dyed (poisoned) bait for control of feral pigs. Hence, the experimental design used was a split-plot design with grain type as a main-plot factor and dyeing as a sub-plot factor. There were no significant differences, as estimated by analysis of variance, in intake of grain types or of dyed and undyed grain.

Naheed & Khan (1989) examined willingness of laboratory rats (*Rattus rattus*) to eat bait. Addition of a low concentration of barium carbonate to bait resulted in significantly reduced intake of the poisoned bait, which the authors called 'poison shyness'. Subsequent removal of the barium carbonate resulted in persistent low intake of the then unpoisoned bait, indicating an acquired 'bait shyness'. Intermittent inclusion of barium carbonate reinforced the poison and bait shyness.

This method has been used for rare non-target species when there are few experimental animals and where mortality of the individuals is to be avoided. Examples of such studies are described in section 3.22.

3.5.3 *Field tests of bait acceptance*

This method is used to assess part (the non-poisoned bait) of the field delivery system, for both target and non-target species. Morgan (1982) described a sequential problem analysis to identify factors that influence bait removal by pest animals in the field. Application of the analysis in New Zealand indicated that over 90% of brushtail possums (*Trichosurus vulpecula*) accepted non-toxic baits in some trials. Acceptance by brushtail possums ranged from a mean of 68% to 100% and varied significantly, as assessed by two-way analysis of variance, between possum populations. There were no significant differences between acceptance of bait types. The analysis was limited as the interaction of the effects of populations and bait types could not be tested because of the lack of replication. Instead, the interaction was used to estimate the residual variance for the *F* ratio tests.

The field acceptance of bait can also be used to evaluate the effects of poisoning. Bamford (1970) used non-toxic flour-paste baits to obtain an index of possum abundance to evaluate the effects of 1080 poisoning.

3.5.4 *Field tests of toxicity*

This method is the field test of toxicity and is based on changes in estimates of pest abundance using methods such as those described in Eberhardt (1978), Caughley (1980), Seber (1982, 1986), Krebs (1989) and Buckland *et al.* (1993) or by following a known number of marked animals, such as by radio-tracking (Amlaner & Macdonald, 1980). Robinson & Wheeler (1983) compared rabbit mortality following control poisoning estimated by spotlight counts with that estimated by radio-tracking. They concluded that spotlight counts were influenced by the behaviour of survivors, whereas the radio-tracking results were not, and spotlight counts should not be used to assess such control because of high variability between counts. The spotlight counts did reveal, however, that immigration was occurring, an event not revealed by radio-tracking.

In many field evaluations of the use of poisons for vertebrate pest control, simultaneous experimental controls (non-poison sites) were not used (Rowley, 1958; Poole, 1963; Batcheler, Darwin & Pracy, 1967; Rowley, 1968; Rennison, 1977; Hone & Pedersen, 1980; Richards, 1981; Rowe, Plant & Bradfield, 1981; Tietjen & Matschke, 1982; Robinson & Wheeler, 1983; Buckle, 1985; Rowe, Bradfield & Swinney, 1985; Balasubramanyam, Christopher & Purushotham, 1985; Hickson, Moller & Garrick, 1986; Knowles, 1986; Thomson, 1986; McIlroy *et al.*, 1986a and King, 1989). In contrast, experimental controls (simultaneous or sequential) were used in field evaluations of poisoning, and some other pest control methods, by Bamford (1970), Cooke (1981), Oliver, Wheeler & Gooding (1982), Hone (1983), Bjorge & Gunson (1985), Foran *et al.* (1985), Crosbie *et al.* (1986), McIlroy, Gifford & Cooper (1986b), Williams *et al.* (1986), Kinnear *et al.* (1988), Cooke & Hunt (1987), McIlroy, Braysher & Saunders (1989), Mutze (1989), Murua & Rodriguez (1989), Choquenot, Kay & Lukins (1990), McIlroy & Gifford (1991) and Twigg, Singleton & Kay (1991).

Mutze (1989) used a factorial design, with poisoning and free-feeding each at two levels, to assess the effect of strychnine on abundance of mice in parts of South Australia. The design was used at each of three sites. The results, as estimated by mice abundance, could have been analysed by analysis of variance, but, surprisingly, were not as the observed results were considered to be 'clear' without statistical analysis.

Field evaluations of poisoning can estimate more than the percentage reduction of pests if two or more independent survey methods are used (Hone, 1983). The survey results can allow three statistics to be estimated: x, the proportion of the pest population that ate the poisoned bait and died; y, the proportion that ate the poisoned bait and survived; and z, the proportion that did not eat the bait.

By definition, the proportional reduction in numbers of pests that ate the bait is:

$$[(x+y)-y]/(x+y) \qquad (3.1)$$

The proportional reduction of the total pest population is:

$$[(x+y+z)-(y+z)]/(x+y+z) \qquad (3.2)$$

Since:

$$x+y+z=1.0 \qquad (3.3)$$

Thus, we have three equations with three unknowns. They can be solved simultaneously to estimate x, y, and z. The model makes several assumptions. Firstly, that following poisoning the survivors have the same countability as before poisoning. If bait-shyness (aversion) occurs then the percentage of pests that ate the bait will be overestimated. Secondly, the model assumes that the population is closed, and thirdly, that pests die only from the primary poisoning and not from secondary poisoning.

Hone (1983) reported the results of 1080-poisoning of a population of feral pigs in southern Australia. The proportional reduction of the feral pig population that ate 1080-poisoned bait was 0.94 and the proportional reduction of the total population was 0.73. Hence:

$$[(x+y)-y]/(x+y)=0.94$$
$$[(x+y+z)-(y+z)]/(x+y+z)=0.73$$

The results estimate the proportion of all pigs that ate the poisoned bait and died (x) was 0.73, the proportion that ate the poisoned bait and survived (y) was 0.04 and the remainder did not eat the poisoned bait ($z=0.23$).

Morgan (1982) described studies of 1080 toxicity to brushtail possums in New Zealand. Instead of using several estimates of population abundance, he used a bait-marking technique that allowed direct estimates of the same parameters. Data in his Table 4 show that for the 1080 concentration of 0.1%, the proportion of the total possum population surviving was 0.32. Hence:

$$y+z=0.32 \tag{3.4}$$

The proportion of the total population that ate the bait was 0.86. Hence:

$$x+y=0.86 \tag{3.5}$$

Substituting equation 3.4 into equation 3.3 gives $z=0.14$. That is, 14% of the total possum population did not eat the poisoned bait. Substituting equation 3.5 into equation 3.3 gives $x=0.68$. That is, 68% of the total possum population died. Therefore, the percentage of the total population that survived even though they ate the poisoned bait was 18%. The results reported by Morgan (1982) can be similarly recalculated for each of his treatments, as shown in Table 3.2.

The above analysis, using either technique, could be usefully applied to studies where changes in bait acceptance are suspected. Such a change has been suggested as a contributing factor to a decline in the effectiveness of 1080 poisoning of rabbits in parts of Western Australia (Oliver *et al*., 1982).

The presence of pesticide resistance can be detected using a test described by Rennison (1977). This is described in section 3.17. The test can also be used to assess the effects of chronic (long-term) pesticides on pest abundance.

3.5.5 *Field tests of sub-lethal effects*

This is the field assessment of sub-lethal effects of poisoning. The assessment commonly uses a survey technique to sample target and non-target animals to determine physiological or anatomical effects. Alternatively, animals may be tested for poison residues. Examples of these studies for non-target species are described in section 3.22.

3.5.6 *Modelling of laboratory or field toxicity*

This is the modelling of toxicity and will be discussed in detail in section 6.3.

3.6 Trapping

Trapping is a widely used method of vertebrate pest control but there have been surprisingly few evaluations of its effectiveness and factors influencing effectiveness. Evaluations have been in the laboratory or field.

3.6.1 *Laboratory test*

Temme & Jackson (1979) described procedures for studying the responses of rodents to traps. Laboratory study of rodent behaviour allows a detailed understanding to be developed of the relative efficiency of different trap designs and emphasises the need when trapping animals cautious to new objects in their environment (neophobic), to place any bait in the most accessible position.

3.6.2 *Field tests*

Fox & Pelton (1977) examined the effect of 10 factors on trap success for feral pigs in Great Smoky Mountains National Park in Tennessee. Trap

Table 3.2. *Estimates of the composition of brushtail possum populations in New Zealand offered sodium monofluoroacetate (1080 poison) or offered no 1080 poison. Estimates were calculated by equations 3.1 to 3.3 using data from Table 4 in Morgan (1982)*

Treatment	Proportion of population		
1080 concentration (%)	Died (x)	Ate bait but survived (y)	Did not eat bait (z)
0	0	0.98	0.02
0.1	0.68	0.18	0.14
0.2	0.83	0.04	0.13

success was estimated as the number of feral pigs captured per trap night. The 10 factors were age of prior pig activity (mainly rooting), location with respect to drainage, site type and forest stand type, daily temperature, temperature deviations, daily precipitation, barometric pressure trends, trap type, and influence of other wildlife species. The only factors that had a significant effect on trap success were age of pig activity and drainage areas. A decline in trap success occurred in areas of older pig activity. The effect of drainage may have been an effect of limited prior control in one area. The evaluation used Chi-square analysis separately for each factor.

A more-efficient evaluation than that above would use all data simultaneously, and test for interactions between factors. Such an analysis could be generalised linear models as used by Saunders (1988). He investigated the effects of factors on the rate of trap bait taken by feral pigs in a part of south-eastern Australia. Each factor, distance to the edge of a patch of trees, distance to road, distance to water, age of pig ground rooting, tracks or dung, and season of the year, was classified at two levels. The analysis showed that each factor, except distance to water, had a significant effect and many factors had significant interactions. The results of the analysis were a set of predictions of the proportion of days in which trap bait was eaten relative to the maximum possible in the experiment.

Factors influencing the likelihood of trapping rabbits (*Oryctolagus cuniculus*) in southern Australia were investigated by Daly (1980). Adult rabbits were most trappable outside the breeding season (late autumn and winter). Trappability increased with age in young rabbits. Age differences were determined by a correlation test and sex differences by a goodness of fit test. Males were more difficult to trap than females during the breeding season.

The trap success of land and raft traps for capture of coypu (*Myocastor coypus*) in Britain was reported by Baker & Clarke (1988). A study site in Norfolk was divided into four blocks with land traps being used in half of each block and raft traps in the other half. Significantly more coypu were trapped in raft than land traps, as assessed by use of the Wilcoxon signed-rank test. There was no significant difference in the number of coypu of different ages and sexes for the two trap types, as assessed by Chi-square analysis. An alternative analysis could be analysis of variance assuming the study used a randomised block design with four blocks and two trapping methods. The raft traps caught significantly fewer non-target species of animals during the test period.

The age structure and sex ratio of coyotes captured in traps and killed with M-44 poisoning devices in Texas were compared by Windberg & Knowlton (1990). There were no significant differences in age structure or sex ratio of coyotes, as assessed by Chi-square analysis. There was also no significant change, as assessed by Chi-square analysis, in the age structure of coyotes trapped during trap and removal sessions over at least 15 days. Choquenot, Kilgour & Lukins (1993) reported a significant bias towards females in feral pigs trapped at a site in south-eastern Australia. The bias was assessed by a Z test of proportions.

The evaluation of trapping for vertebrate pest control has seemingly made poor use of the theory of capture success that has been developed for population estimation. Mark–recapture studies repeatedly indicate that the probability of capture is not equal for all animals in a population (Caughley, 1980). Models of mark–recapture for population estimation allowing for three sources of unequal capture probabilities have been described (Otis *et al.*, 1978, Pollock *et al.*, 1990). Capture probabilities vary with: (i) time of trapping occasion, (ii) behavioural responses to trapping, and (iii) features such as age or sex of individual animals. Presumably (ii) is not important for animals that are killed after trapping but could be important for animals that escape from traps.

Skalski, Robson & Matsuzaki (1983) and Skalski & Robson (1992) discussed the relationship between trapping effort and catch per unit effort (CPUE). If traps were distributed at random with respect to the home range of the animals then with increasing trapping effort more animals would be captured. However, as the study area became saturated with traps then CPUE should decline (Fig. 3.4). That prediction was tested against the hypothesis that CPUE was constant for all levels of the number of traps (Fig. 3.4). The relationships were tested by experimental trapping of small mammals in Washington state which resulted in rejection of the second (constant) and acceptance of the first (declining) relationship. Gosling & Baker (1987) reported a linear increase in catch success (numbers of female coypu trapped / total numbers of female coypu / trapper) with increasing trapping effort (mean

Fig. 3.4. Hypothetical relationships between the catch per unit effort (CPUE = animals captured per trap) and the number of traps in an area as described in the text. (After Skalski *et al.*, 1983.)

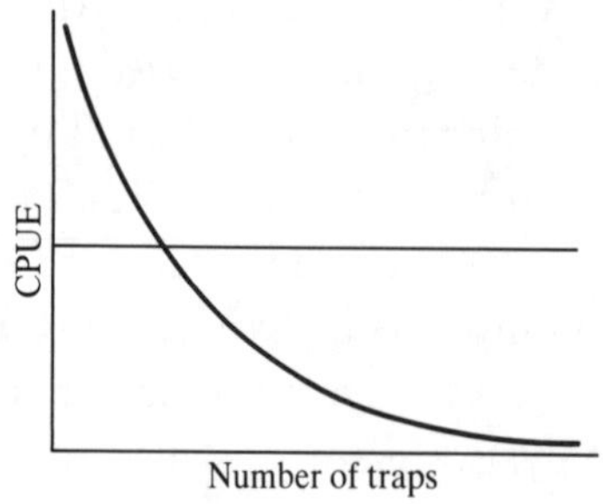

number of traps set / trapper / night of trapping). The linear trend was possibly the result of trappers not setting too many traps close together.

The analysis of the effects of trapping pests on islands presents interesting problems. Statistics on the number of pests killed and trapping effort can be easily documented, as Fitzgerald & Veitch (1985) have done for trapping cats on Herekopare Island in New Zealand. On Herekopare Island the short-term effects on birdlife of trapping were not assessed. Cats may have previously had dramatic effects on sea and landbirds but the effects of removing the cats were not estimated. That would require extensive survey work on the seabirds and landbirds, ideally before and after the cat removal. Another common problem in interpreting the effects of such pest control is that islands are studied one at a time so there is no replication and no experimental control.

3.7 Fencing

Fencing is used to reduce or stop immigration of pests onto crops and other areas of commercial or conservation interest. The trend in recent years has been to evaluate electrified fences because of their potential repellency compared with non-electrified fences. The use of fencing has been carried to extremes in Australia with thousands of kilometres of fencing on land to exclude dingoes and in the ocean to exclude sharks from beaches (Woods, 1974). Fencing is also used widely elsewhere, such as in Zimbabwe to control wildlife movements as part of an attempt at control of foot and mouth disease (Taylor & Martin, 1987). An evaluation of the efficiency of fencing in such a disease-control project would be very interesting. McKillop & Sibly (1988) recommended that research on fencing, especially electric fencing, take greater account of animal behaviour to improve effectiveness. Three methods are used to evaluate fencing and these are now described.

3.7.1 *Laboratory*

Shumake *et al.* (1979) observed the fence-climbing behaviour of Phillippine rice-field rats (*Rattus rattus mindanensis*) in test chambers in a laboratory. Three fence designs were evaluated. The fences were later used in semi-field situations inside a coverted grain silo.

3.7.2 *Semi-field tests*

A small-paddock evaluation was used by Hone & Atkinson (1983) to test the efficiency of electrified and non-electrified fences in limiting movements of feral pigs in Australia. Paddocks 30 m by 20 m were used, with the test fence being one of the 20 m sides of the paddock. The experimental design was a randomised complete block design. An experimental control, such as an unfenced side, was not used. The non-electrified fences were used first then all were electrified, hence the non-electrified fences were a sequential experimental control for the electrified fences. The pigs rapidly became conditioned to the electric fences as contacts rapidly decreased within one hour of exposure. Efficiency was evaluated by analysis of variance of three variables – the percentage of pigs that crossed a fence type, the time till the first pig crossed and the cumulative time that pigs stayed in the test paddocks behind each fence. The experimental approach allowed a thorough evaluation of many fence designs but is limited by the need to extrapolate from the small paddock to field situations.

An electric fence was evaluated by Porter (1983) by comparison of new growth on apple trees inside and outside a fenced experimental area in New York state. The design of paired exclosure and open sites of 1 ha or 5 ha was replicated. The statistical test used to assess the significance of the results was not described in the paper.

3.7.3 *Field tests*

A brief account of a field test was given by Forster (1975). The number of sandwich terns (*Sterna sandvicensis*) nesting at a nature reserve in Aberdeenshire, Scotland, increased when an electric fence was erected to limit immigration by foxes (*Vulpes vulpes*). The effect of the fence was assessed by an unanalysed comparison of the number of pairs of terns prior to (the year before) and after the fence construction. This is an example of an unreplicated sequential experimental control. Shaughnessy (1980) reported a decline in a population of the rare jackass penguin (*Spheniscus demersus*) on Sinclair Island off Namibia. The decline followed changes to a concrete fence that had been built to protect a nesting penguin colony from predation by Cape fur seals (*Arctocephalus pusillus*). The population of jackass penguins on nearby islands that did not have seals did not decline as dramatically as on Sinclair Island. Hence, this evaluation of a fence had at least one simultaneous experimental control.

Shumake *et al.* (1979) also used a field test of electric fences to reduce rat damage to rice crops in the Philippines. Paired sites were used to compare two fence types. Data were analysed as a factorial design (fence types, sites, repeated in time) by analysis of variance. This showed a significant difference in rat activity inside and outside the fences. Lokemoen *et al.* (1982) reported the effects of electric fences on mammalian predation of eggs and nestlings in waterfowl nests in North Dakota and western Minnesota. For the evaluation, paired (fenced, unfenced) sites were used in or near wetlands. Nest success averaged across nine paired sites was higher inside (unweighted mean = 59%) the fenced areas than outside them (unweighted mean = 23%). The statistical significance of the difference was not tested. Nests were still disturbed inside the fenced areas but they were thought to be by avian predators or a few mammals that passed under or through the electric fence. A similar paired experimental design was used by Hygnstrom & Craven (1988) to test the effects of fencing to reduce deer (*Odocoileus* spp.) damage to cornfields in Wisconsin. Paired sites were tested for their differences in variables of area, perimeter/area ratio, plant density, and percentage and duration of field exposure to deer damage. Differences between fenced and unfenced sites were compared by Tukey's HSD test, which showed a highly significant difference. Cornfields inside the fences had much lower damage than unfenced fields.

The effects of fence types on damage to lupin crops by western grey kangaroos (*Macropus fuliginosus*) in south-western Western Australia were reported by Arnold, Steven & Weeldenburg (1989). Four fence types were tested in each of two years at each of two sites in a factorial design. The grain yields of lupins were compared with those of fully protected crops. The effects of fence type, as assessed by analysis of variance, differed between years and sites. The ringlock (netting) fence, topped by barb wire, gave the greatest protection.

3.8 Aversive conditioning

The reduction of predation by conditioned taste aversion in predators has been examined in many studies. This section will give a short overview of selected studies. The hypothesis is that predators that find and eat a bait treated with a strong emetic, such as lithium chloride (LiCl), will become ill and subsequently avoid eating live prey animals. The coyote has been the main predator studied and sheep the main prey. Two methods have been used in evaluating the aversive conditioning: pen and field tests.

3.8.1 *Pen tests*

Gustavson *et al.* (1974) used individual coyotes in pens to study aversion to lithium chloride. Aversion to eating hamburger was established following prior ingestion of hamburger laced with lithium chloride. There were no coyotes that were experimental controls, that is, simply offered hamburger on both test occasions, and offered safe dogfood on other days. Subsequently, testing of coyotes showed that exposure to lithium chloride in lamb flesh stopped one of three coyotes from killing a lamb. Rebaiting and intraperitoneal injection of the other coyotes then stopped their killing of lambs on a subsequent test. No coyotes were used as experimental controls, that is, baited and injected but not injected with lithium chloride.

Conover, Francik & Miller (1977) reported that coyotes did not stop killing and eating live chickens when presented in random sequence with treated dead chickens. The coyotes ate at least some or all of the treated dead chickens. A second experiment with live and dead mice and intraperitoneal injection of lithium chloride after eating dead mice reached the same conclusions; the coyotes did not generalise from the treatment or treated bait to the live animals. Conover *et al.* (1977) noted that since the experiments used small sample sizes and did not use sheep as prey then the results had limited relevance to predation of sheep by coyotes. The results of Conover *et al.* (1977) were disputed by Gustavson (1979) but not satisfactorily according to Conover, Francik & Miller (1979).

Horn (1983) reported that an intraperitoneal injection of a lithium chloride solution to coyotes did not stop killing. Each coyote had previously killed prey (domestic rabbits *Oryctolagus cuniculus*). Injection resulted in sickness but all coyotes subsequently killed rabbits. Horn (1983) also described a different experiment. Coyotes killed rabbits and subsequently were offered and fed on rabbit meat packages. Initially, the meat packages contained empty gelatin capsules and subsequentially the coyotes killed rabbits. Later, the same coyotes were offered meat packages in which the capsules contained lithium chloride. Hence, the empty capsules were used as experimental controls. The coyotes subsequently killed rabbits after ingesting the treated baits. The treatment phase and tests with live rabbits were repeated with the same results. In another experiment, coyotes were successively given a bait package with lithium chloride, a carcass laced with lithium chloride and received an intraperitoneal injection of lithium chloride. Aversions to live prey were subsequentially recorded.

In an experiment by Burns (1983) the transfer of bait aversion to prey-killing aversion was tested by comparing the prey-killing behaviour of a test group of coyotes (previously fed lithium chloride baits) with that of a naive group (no lithium chloride baits). The time taken to kill prey (a live sheep) was 2.7 days for both the treatment and control groups of coyotes indicating no transfer of bait aversion to prey-killing aversion.

3.8.2 *Field tests*

Bourne & Dorrance (1982) determined whether aversion would influence sheep predation. A comparison of replicated farms in southern Canada over three years with and without lithium chloride treatment found no significant effect of bait type (placebo versus lithium chloride) or years or their interaction, as assessed by analysis of variance.

Conover (1990) reported that sequential presentation of treated and untreated birds' eggs significantly reduced egg predation by mammals. The treated eggs were injected with emetine dihydrochloride in field tests in southern Connecticut. The predators were primarily raccoons (*Procyon lotor*), opposums (*Didelphis virginiana*) and striped skunks (*Mephitis mephitis*). In contrast, during simultaneous presentation of treated and untreated eggs the predation rates were not significantly different. It was concluded that it might be possible to reduce predation of eggs by mammals if treated eggs are distributed prior to, but not after, the commencement of egg laying.

3.9 Chemical repellents

Laboratory tests were used by Sullivan, Crump & Sullivan (1988b) to estimate the effect of predator odours on the number of captures of northern pocket gophers (*Thomomys talpoides*). The comparisons used control and treatment rooms alternated between tests. Gophers significantly avoided a compound from fox faeces, as assessed by Chi-square analysis, in two of three tests.

Winter field tests in British Columbia of predator odours including a synthetic fox urine mixture, were reported to significantly reduce vole (*Microtus montanus* and *M. pennsylvanicus*) attack on apple trees (Sullivan *et al.*, 1988a). The effect varied, however, with different odours, including no effect at all. All tests used control and treatment blocks simultaneously and results were analysed by Chi-square analysis.

Hygnstrom & Craven (1988) compared two chemical repellents in their effect on deer (*Odocoileus* spp.) damage to corn in Wisconsin. Replicated corn fields were randomly allocated treatments, though with some practical limitations because of local conditions, and Tukey's HSD test was used to determine differences between repellents and experimental controls. Estimated yield losses were higher on treated than experimental control fields.

3.10 Sonic devices

Woods (1974) considered that sonic devices were of little use for pest control because of habituation. He also noted that there had been little scientific evaluation of the devices.

Bomford (1990a) tested the effectiveness of a sonic device for deterring starlings (*Sturnus vulgaris*) in a part of southern Australia. The sonic device was directed at three segments of a circular plot and three untreated segments were screened from the device in a factorial design. The effects of the device on starling numbers were estimated by three-way analysis of variance with the device, time (pre-treatment and post-treatment) and distance (near and far) as the three treatments. There were no significant differences in starling numbers between segments with and without the device, or at near and far distances.

Starling numbers were significantly higher when the device was operating than when it was not.

A review of the evaluations of many sonic devices used in vertebrate pest control was reported by Bomford & O'Brien (1990). They concluded that many evaluations lacked experimental controls or true replication and the interpretation of results was often limited by potential confounding of treatments and other effects.

3.11 Biological control

Biological control is the use of living organisms as pest control agents (Waage & Greathead, 1988). The agents include parasites, parasitoids, predators and competitors. Begon *et al.* (1990) described four types of biological control: (i) introduction of a natural enemy, such as a predator, parasite or pathogen, (ii) inoculation involving periodic release, (iii) augmentation involving the release of a natural enemy already present, and (iv) inundation involving mass release. The first type is the classical method of biological control and will be the method emphasised here.

There are few successful examples of the biological control of vertebrate pests (Davis, Myers & Hoy, 1976). Those available use pathogens as the biological control agent. Early unsuccessful attempts include use of predators such as mongoose (*Herpestes auropunctatus*) to control rats in Hawaii and common mynas (*Acridotheres tristis*) to control cane beetles in northern Australia.

Krebs (1985) described five attributes of biological control agents that are important, but not sufficient, for successful control: (i) general adaptation to the host and environment, (ii) high searching capacity, (iii) high rate of increase relative to the host, (iv) general mobility adequate for dispersal, and (v) minimal lag time in responding to changes in host abundance. The analysis of Murdoch, Chesson & Chesson (1985) suggests that these attributes are not all necessary for successful biological control. They added that a stable pest equilibrium is not a necessary condition for control.

Two population processes have been associated with models of successful biological control: depression of pest abundance, and maintenance at a lower level (Waage & Greathead, 1988). The depression characteristic has not been challenged but the stability criterion has (Murdoch *et al.*, 1985). Waage & Greathead (1988) concluded that maybe the desirable attributes of control agents varied with pest systems so no overall set of desirable attributes could be described. The analysis of biological control of vertebrate pests has concentrated on the evaluation of the depression of abundance rather than the maintenance of the lower level. Spratt (1990) listed eight characteristics of an ideal biological control agent, though surprisingly there was no comparative analysis of lists, such as with that of Krebs (1985). Obviously, further research is needed to develop a better list or clearly show there is no best list.

The theoretical bases of biological control of pests, particularly insect pests, were reviewed by Murdoch *et al.* (1985). They concluded that there was, at that time, no general theory that explained the successes of biological control. This agrees with the view of Krebs (1985) that biological control is akin to gambling; it is sometimes successful. The process of selection of likely biological control agents was discussed by Dobson (1988). Introduced animals on islands can have a limited range of pathogens and parasites compared with the same species on a mainland. The missing pathogens or parasites may be candidates for use in biological control. Freeland (1990) suggested a similar idea for control of large introduced herbivores in northern Australia. The theoretical basis of biological control was further examined by Anderson (1982a) and is discussed in section 6.4.

Three methods have been used for evaluating biological control of vertebrate pests: laboratory assessment of specificity and effects, field assessment of lethal effects, and modelling.

3.11.1 *Laboratory tests*

Prior to release of a potential biological control agent it is necessary to check if the agent already occurs in the test population. The evaluation of a potential biological control agent may require estimating the specificity of a likely vector. That is, will the potential vector transmit the pathogen between hosts and will non-target species be infected by the potential vector and if so will the non-target individuals develop the disease? The evaluation of potential vectors requires extensive laboratory and field experiments. The evaluation of *Capillaria hepatica* as a biological control agent of house mouse in Australia includes an assessment of the likely non-target species that could be infected and the impact on their populations (Spratt, 1990). Details of a field assessment were provided by Singleton *et al.* (1991b).

The effects of a potential biological agent can be examined in laboratory tests. Singleton & Spratt (1986) studied the effects of *Capillaria hepatica* on fecundity in laboratory mice and reported a significantly lower number of young weaned per female infected mouse compared with uninfected female mice, as analysed by Student's *t* test.

3.11.2 *Field tests*

Myxomatosis virus was introduced to Australia in the 1940s and 1950s to control rabbits. Early research in the 1940s in South Australia and in 1950 in Victoria and New South Wales gave discouraging results and it was thought that the virus may be of limited value for rabbit control (Fenner & Ratcliffe, 1965; Fenner, 1983). In late 1950 and early 1951 the virus apparently had a devastating effect on rabbit populations, though there is little quantitative data to support this (Fenner, 1983).

Myers (1954) reported results from four study sites of the effects of myxoma virus introduction. Enough quantitative data were reported from one study site to show a series of damped oscillations (decreasing amplitude) in the number of carcasses of rabbits that presumably died from the disease. The disease disappeared despite an abundance of rabbits. It was concluded that the one factor lacking was transmission by flying insects. Myers, Marshall & Fenner (1954) reported a spectacular decline in counts of healthy rabbits on one standard transect in an area at Lake Urana (New South Wales) shortly after deliberate introduction of myxomatosis. The rabbit population was believed to consist solely of susceptible individuals. However, no trends were reported from areas without myxomatosis. Serological evidence showed that 76% of rabbits that were still alive after the epizootic had not been infected, hence 24% had been infected but had survived. Subsequent epizootics (epidemics in wild animals) produced lower mortality rates in rabbits because of attenuation (lower virulence) of the virus. Mosquitoes were identified as important vectors of the pathogen, and probably necessary for the maintenance of infection.

Myers (1970) stated there were no comprehensive data sets to describe the effect of myxomatosis on the rabbit population in Australia. He provided two new data sets that showed that exports from Australia of rabbit skins dropped sharply after introduction of myxomatosis, and indices of rabbit abundance were lower in 1970 than 1950 at four sites in southern Australia. There were no data from comparable sites where myxomatosis did not occur.

Cooke (1983) investigated the role of the European rabbit flea (*Spilopsyllus cuniculi*) as a vector of myxoma virus. One population of rabbits was the experimental control, one population had fleas introduced but they were not infected with myxoma virus and the third population had fleas and myxoma virus introduced (as infected fleas). The response was estimated as the proportion of the initial population that survived, which was lowest (0.1) for the population which received infected fleas. The intermediate response (0.4) of the population which received non-infected fleas was associated with mortality from a field strain of myxomatosis, the effects of which could be distinguished from that of the experimental myxomatosis. The proportional survival at the experimental control site where no fleas and no myxoma virus were introduced was 0.8. The statistical significance of the results was not tested.

Ross *et al.* (1989) distinguished between two estimates of mortality in assessing the effects of myxomatosis on rabbits in England and Wales. The first was case fatality, the proportion of infected hosts that died from the disease. The second was field mortality rate, the proportion of the total host population that died from the disease. Ross *et al.* (1989) estimated that case fatality rates ranged from 47–69% and field mortality rates ranged from 12–19% per outbreak of myxomatosis.

Jaksic & Yanez (1983) discussed the introduction and control of European rabbits (*Oryctolagus cuniculus*) on Tierra del Fuego. It was considered that rabbits had increased in abundance since introduction in 1936. Foxes (*Dusicyon griseus*) were introduced in 1951, and myxomatosis in 1954 as biological control agents. Analysis of fox diet suggested that the foxes had not and could not control the rabbit population. The study concluded that myxomatosis had greatly reduced the rabbit population though, surprisingly, no quantitative data were given to support the conclusion.

Feline parvo virus was introduced to Marion Island in the southern Indian Ocean in 1977 to control feral cats (*Felis catus*) and their effects on nesting seabirds (van Rensburg, Skinner & van Aarde, 1987). The virus causes the disease feline panleucopaenia (FPL) in cats. The disease is host-specific, highly contagious and causes high mortality amongst kittens. The cat population on Marion Island dropped from an estimated 3409 in 1977 to 615 in 1982, suggesting an annual rate of decrease of 29%. The litter size decreased, age structure changed and mortality rates were higher in 1982. There was evidence that the cat population was declining at a slower rate in 1982 than previously (van Rensburg *et al.*, 1987). This could have been associated with decreased spread or virulence of the virus, as occurred with myxomatosis and rabbits in Australia. The evaluation of the effect of biological control occurred without a simultaneous experimental control. That could only have been obtained if a nearby island had a similar cat population. Alternatively, a sequential control could have been obtained by having more estimates of population size prior to the introduction of the virus. Despite these apparent experimental limitations to the evaluation, the trend in the cat population follows that expected from a knowledge of the disease. There were no data on the effect of pathogen introduction on populations of seabirds.

Nettles *et al.* (1989) reported use of swine fever (hog cholera) on islands off California for biological control of feral pigs. The abundance of feral pigs apparently dropped rapidly but the disease did not persist, though the pigs apparently did. That may have been because the abundance of feral pigs, after disease introduction, was reduced to below the threshold host abundance for disease persistence, or because all the susceptible pigs that survived became immune. The evaluation was limited because of the lack of accurate or precise data on abundance of feral pigs before or after disease introduction. The more detailed report by Nettles *et al.* (1989) appears to contradict the statement by Davis *et al.* (1976) that hog cholera eradicated wild pigs.

3.11.3 *Modelling*

Details of modelling work are given in section 6.4 and related issues in section 6.8.

3.12 Shooting

Shooting has been used for vertebrate pest control for many years, but with limited evaluation of its effectiveness. The shooting is usually ground-based but can also occur from a helicopter or light aircraft. Evaluation has been by two methods: field tests and modelling.

3.12.1 *Field tests*

McIlroy & Saillard (1989) described an evaluation of hunting, with dogs, of feral pigs in mountain forests of south-eastern Australia. The hunters killed only 13% of pigs known to be in the area. Radio-tracking was used to estimate home range size before, during and after hunting, in two study sites. There was a significant interaction, as assessed by analysis of covariance, of home range size between sites and time periods. The average home range size increased after hunting at one site but not the other. The number of locations (fixes) was used as a covariate in the analysis, as the estimated home range size was dependent on the number of locations. The nearest distances that hunters approached each radio-collared pig were also estimated for each of the five days of the experiment. The average distance was significantly higher, as assessed by analysis of variance, on the fifth day.

The longer-term effects of shooting can be determined by estimating the rate of increase of the pest population after shooting. Choquenot (1990) reported the trend, after shooting, of a population of feral donkeys (*Equus asinus*) in northern Australia. The observed instantaneous, also called exponential, rate of increase of the population was 0.21/year. The rate of increase of a donkey population that had not been subject to shooting was not reported, though Freeland & Choquenot (1990) and Choquenot (1991) reported that a nearby donkey population not subject to shooting had not increased significantly over the same time period.

The eradication, by shooting, of feral goats (*Capra hircus*) off Raoul Island in the southern Pacific Ocean was described by Parkes (1990). The operation was monitored by trends in the number of goats killed and the number killed per hunter per day. The same sort of data were used to evaluate trends in goat numbers in other parts of New Zealand in relation to shooting pressure. For example, the number of goats killed per day hunted declined significantly ($r^2 = 0.81$, $P < 0.001$) from 1961 to 1987 in Mt Egmont National Park but reanalysis of that and additional data showed that from 1981 to 1989 the number of goats killed per day hunted had not changed significantly ($r^2 < 0.01$, $P > 0.05$).

A comparison of the efficiency of shooting buffalo (*Bubalus bubalis*) in northern Australia either from the ground or from a helicopter was reported by Boulton & Freeland (1991). The predicted number shot per day was related by linear regression to population density at the start of the day. Regressions were estimated separately for each shooting method and each was highly significant. The slopes and intercepts of the regression lines could have been compared in an analysis of covariance, to test for differences. Given the statistical significance of the fitted regressions each would probably have been different.

3.12.2 *Modelling*

Shooting, from a helicopter, of feral pigs, buffalo and feral donkeys, has been evaluated by estimating the relationships between the duration of shooting, the initial abundance of pests, and the number of pests shot. The evaluation, by modelling the shooting process, is described in section 6.5.

3.13 Chemosterilants

The use of chemicals to sterilise all or part of a pest population has been studied in many areas for many pests. A detailed evaluation of the studies is not reported here. Bomford (1990b) reviewed many studies of fertility control including chemosterilants and reported that they usually lacked adequate experimental controls, replication and detailed analysis. Murton, Thearle & Thompson (1972) had earlier expressed some of the same conclusion on studies of the effects of chemosterilants on abundance of feral pigeons (*Columba livia*), as well as expressing concern about biased conclusions because of biased methods of estimating population size of the pigeons. Despite these concerns, evidence on the effects of some pesticides on falcons in northern America and Europe (Newton, 1979) indicates population reduction can occur because of reductions in fertility.

3.14 Multiple evaluations

There have been few simultaneous evaluations of multiple control methods. This is probably to be expected given the logistic difficulty of doing so. Two well-designed such experimental evaluations have been reported for rabbit control in South Australia (Cooke, 1981; Cooke & Hunt, 1987) and one in central Australia (Foran *et al.*, 1985). The designs and analyses are described in section 3.20. Evaluation of several control methods simultaneously has the advantage of being able to test for interactions between the effects of methods and compare costs and benefits more directly.

Murua & Rodrigeuz (1989) examined the effects of habitat manipulation and poisoning on rodent damage to pine plantations in Chile. The factor, habitat manipulation, was at three levels, pruning, pruning and thinning of the pine trees and an experimental control (no pruning or thinning). The three levels of the factor, habitat manipulation, corresponded to three study sites, hence sites and levels were confounded. Within each study site three levels of poisoning were used: no poisoning, brodifacoum and bromadiolone. The experimental design was therefore a split-plot design with habitat manipulation as the main-plot factor, and poisoning as the sub-plot factor with a total of nine treatment combinations. The data were analysed, however, as a factorial design with both main- and sub-plot factors being tested on the same residual variance. The interaction of main- and sub-plot factors was not tested. Murua & Rodrigeuz (1989) reported that rodent damage was reduced more when both poisoning and pruning and thinning occurred.

3.15 Predation control

The methods and statistical analyses used to study predation of livestock have been discussed in section 2.4. The focus here is on the analysis of predation control.

3.15.1 *Problem definition and species*

Research on methods of control of coyote predation was reviewed by Sterner & Shumake (1978). They provided a comprehensive assessment of evaluations of methods up to that time. It was noted that the evaluations were often difficult because of movement of coyotes into and out of study areas and variability in response variables – the measures used to assess effectiveness of methods. Wade (1978) also reviewed methods of predation control, though the emphasis was on the history and administration of control. Research on predation control since those reviews will be described here, though this is not intended to be exhaustive.

Murton (1968) suggested that attacks on lambs by ravens and crows in highland Britain may be related to the social structure of the bird populations. Dominant birds in both species occupy the best territories and subordinates occupy marginal habitats, such as around flocks of lambing sheep. As a consequence, control of lamb attacks should be concentrated around lambing flocks and not be directed against all ravens and crows. The idea appears not to have been tested. Several predators kill more prey than they eat – surplus killing. O'Gara *et al.* (1983) reported this for coyotes, and Rowley (1970) reported it for foxes. Hence, a small amount of selective control may yield a large response in damage reduction. This idea also appears to be untested.

3.15.2 *Methods and analysis*

The methods of predation control can be broadly classified into lethal and non-lethal and the methods of study into inferential and experimental. Inferential studies are those based on observation of situations but where there is no experimental control or manipulation. Both will be discussed briefly.

An inferential study occurred in the western U.S.A. where a common method of predation control was use of sodium monofluoroacetate (1080). Up till its ban as a predacide on federal lands in 1972, there had been a gradual decline in its use in some areas. Lynch & Nass (1981) reported that there was a significant negative correlation between the number of 1080 bait stations used and the reported losses of sheep and goats in western national forests from 1960 to 1972. As 1080 use declined predation losses increased. The estimates of losses represent the difference in the number of animals released into the forest each summer, and the number removed in autumn (fall). Hence, the difference is caused by predation, weather, toxic plants and other causes, as discussed by Lynch & Nass (1981). Similarly, the number of 1080 bait stations used is not necessarily a good measure of 1080 actually eaten by predators. A lot of 1080 may have been offered but not eaten.

Non-lethal methods of predation control were re-examined after Gustavson *et al.* (1974) reported experimental studies of conditioned aversions of coyotes. The details of the methods of study were described earlier in section 3.8. This study generated debate with most controversy centered around whether the learned aversion can be generalised to live prey, experimental procedures used (Conover *et al.*, 1977; Gustavson, 1979; Conover *et al.*, 1979), and the disparate results of field studies (Gustavson *et al.*, 1976; Bourne & Dorrance, 1982). The continued debate suggests that the method has not been adequately tested; there has been limited use of experimental controls and inadequate replication. I suggest these same conclusions can be applied to many topics in vertebrate pest control.

Green & Woodruff (1983) described two methods of evaluating the effectiveness of guard dogs as a method of predation control. The first method used simultaneous comparisons of predation in sheep flocks with and without guard dogs. The second method used sequential, between time periods, comparisons, again with and without guard dogs. Green & Woodruff (1983) considered that neither method was free of bias resulting from factors affecting predation. That view is incorrect if study flocks were selected at random, and flocks replicated, as then the effects of such factors would be expected to be similar in the two sets of flocks. The results were described of 12 trials which used one or the other method of evaluation. In five trials the number of sheep killed by predators was considered to be appreciably less when guard dogs were present. However, no statistical analysis of comparative mortality in any trial, or assessed across all trials, was reported. Despite this it was considered that the use of guard dogs was a practical method to protect sheep on rangelands. Further, more critical, evaluation of the method is required.

Bjorge & Gunson (1985) reported that total mortality of cattle dropped significantly after wolf control using strychnine in Alberta. The number of cattle mauled by wolves was also significantly lower after control. The number of cattle killed by wolves also dropped but was not significantly lower. This study estimated the response of predation to control by use of a sequential experimental control and tested statistical significance using Chi-square analysis.

Taylor (1984) concluded that experimental studies of predation were more revealing than descriptive or inferential studies. He, however, argued that experimental studies should include not only treatments where predator abundance is reduced, but also treatments where prey abundance is reduced, to increase understanding of the effects of predators on prey abundance and stability. The latter study (prey reduction) could test the hypothesis that exotic predators (such as foxes in Australia) may switch prey (for example to native endangered species) when control of some other prey occurs (such as rabbits).

The economic analysis of predation control is discussed in section 4.4 and modelling in sections 5.5 and 6.7.

3.16 Control of infectious diseases

Statistical analysis of damage by infectious diseases was discussed in section 2.5.

3.16.1 *Problem definition*

The analysis of control of infectious diseases in wildlife involves several steps. Firstly, a clear distinction must be made between the wildlife as pests and the wildlife as hosts of the pathogen. Secondly, a distinction must be made between control of the wildlife and control of the pathogen or disease.

Anderson (1984) noted several key points to achieve control of diseases. Firstly, an understanding of the dynamics of the disease, rather than a simple statement of prevalence levels, is needed. If the disease is stable to disturbance then control will need to be continual. If, however, the disease system is locally stable then periodic control may reduce disease incidence to another locally stable state. Secondly, the concept of a threshold host density suggests an objective different to host eradication, that is, pathogen eradication by host population control. Thirdly, if vaccination is being considered then modelling can estimate the proportion of hosts that need to be vaccinated. This is described in more detail in section 6.8.

Greenhalgh (1986) discussed several objectives of control of infectious diseases in humans. The differences between the objectives appear trivial but on closer examination require such differentiation. The results should apply equally to pest animals and infectious diseases spread by them. When control was by vaccination then as many susceptible animals should be vaccinated as possible whether the aim is to maximise the number vaccinated or minimise the number infected. If control was by isolation of animals to stop spread of infection, then isolation of the most infectious individuals would minimise the number of animals ever to be infected. Alternatively, if the aim is to maximise the expected number of animals alive at some future time, then a strategy of switching from no removal to full removal effort of the least infectious individuals is needed. The results, obtained by modelling, have not been tested empirically. The different objectives would require different evaluations, emphasising either the temporal trends in infection or host abundance at pre-determined future times. Smith & Harris (1991) showed, by modelling, that fox control could reduce the number of foxes infected with rabies but increase the duration of rabies outbreak.

3.16.2 *Methods and analysis*

The analysis of response to disease control has been based on laboratory and field evaluations. Hypotheses to test in the evaluation of control of infectious diseases in wildlife are listed in Table 3.3. The concept of the threshold host abundance is described in more detail in section 5.6. Analysis of control of infectious diseases has used three methods: laboratory, field studies and modelling. Studies of modelling of control of infectious diseases are described in section 6.8 and related studies of biological control using infectious diseases are described in section 6.4.

Laboratory analysis of disease control has emphasised the disease susceptibility of species, pathogen survival and spread and vaccine evaluation. The vaccination of foxes against rabies has been studied in the laboratory and in the field. Blancou *et al.* (1986) reported the results of laboratory tests of oral vaccination of foxes with a live recombinant vaccinia virus. Administration of the vaccinia virus elicited an antibody response that protected all 12 foxes against later exposure to rabies virus. All six unvaccinated foxes that were exposed to rabies virus died. When each vaccinated fox was placed in a pen with one other unvaccinated fox then three of four unvaccinated foxes died and the fourth survived. The latter event was suggested to be an artifact of accidental contact of the unvaccinated fox with some of the vaccine on the mouth of the vaccinated fox. Brochier *et al.* (1988a) also reported a laboratory evaluation of a rabies vaccine for foxes.

Analysis of field studies of disease control has emphasised the occurrence of disease in wildlife with little study of the effects of wildlife disease control on disease prevalence or incidence in livestock or humans. Macdonald (1980) described the results of efforts at rabies control in Britain in the nineteenth century. Then rabies was present in Britain and mostly reported in dogs. Legislation was passed that required the muzzling of dogs, but apparently that had little effect as people did not do it. Then stray dogs were allowed to be seized and a strict quarantine period imposed on imported dogs. Rabies then disappeared. Wildlife appeared to have played little if any role in maintenance of rabies. In Europe vaccination of foxes against rabies occurs in May/June and/or September/October (Brochier *et al.*, 1988a). The vaccination programme has had little detailed evaluation, though early results were favourable (Bacon & Macdonald, 1980).

A field test in Belgium of fox vaccination using a recombinant virus reported that after 15 days 56% of baits and capsules (that contained the vaccine) had disappeared (Pastoret *et al.*, 1988). Foxes were reported to be numerous in the area. No data were presented on whether foxes ate any of the vaccine bait, whether any foxes seroconverted and hence whether any foxes were vaccinated against rabies. Two of nine species of small mammals in the area were identified as having eaten some vaccine bait, based on the presence of tetracycline marker in the animals.

Table 3.3. *Testable hypotheses for which data are needed to evaluate the effects of disease control in wildlife*

1. Control of wildlife population reduces disease prevalence (proportion of host population infected) in human, livestock or wildlife hosts
2. Maintenance of the wildlife population below a specified level of abundance or density (the threshold) is associated with disease disappearance or reduced prevalence in human, livestock or wildlife hosts
3. Control of potential disease vectors reduces the disease prevalence in human, livestock or wildlife hosts

In a more detailed analysis, Brochier *et al.* (1988b) compared the number of rabies cases in animals (wild and domestic) in a part of Belgium before and after a vaccination campaign in both a treated (vaccination occurred) and an untreated (no vaccination occurred) area. Hence, the analysis had sequential and simultaneous experimental controls but only temporal replication. The number of rabies cases was lower in the treated area after vaccination but in the untreated area was similar to the year before. There was no statistical analysis. The vaccine baits were marked with a tetracycline marker which could be detected on post-mortem. Analysis showed that several foxes were rabid yet had eaten the vaccine baits as evidenced by presence of the marker in the jaw bone. Brochier *et al.* (1988b) suggested that this may have occurred because the foxes were already infected when they ate the vaccine bait or had not broken the capsule when they ate the bait so had not been exposed to the vaccine.

Blancou *et al.* (1988) described use of a vaccine (SAD B19) in Europe. The results were assessed by the rate of bait removal and evidence of seroconversion (presence of antibodies in the blood) in baited areas. This implies there was no assessment of the incidence of rabies or comparison with areas not baited.

MacKenzie (1990) reported, in general terms, that results of field vaccination of foxes were encouraging and Smith & Harris (1989) reported in similar general terms that culling of foxes had controlled rabies in France, Denmark and Austria. Macdonald (1980) cast some doubt on the results in Denmark by noting that rabies disappeared in an area in Germany at the same time, without any attempts at control.

General descriptions of rabies control in Switzerland were reported by Macdonald (1980). The incidence of rabid foxes was lowest at a time of gassing campaigns, but the incidence increased during the later stages of the campaign and further increased after cessation of gassing. The increase in incidence would be expected after control ceased. Whether the low level of rabies was a result solely of the control is difficult to assess as there were no data on incidence in an area where no control occurred. Similarly, there were no data on the effect of gassing foxes on the incidence of rabies in humans.

The incidence of rabies in striped skunks (*Mephitis mephitis*) in Saskatchewan was higher than that in Alberta and Montana (Pybus, 1988). The difference was thought to be associated with more skunk control, by poisoning and trapping, in Alberta and Montana, though the data were not statistically analysed.

Anderson & Trewhella (1985) reported a decline in the prevalence of tuberculosis (TB) in badgers in three counties of England from 1971 to 1982 inclusive. The decline occurred at the same time as control of badgers occurred, which apparently reduced badger populations. It was inferred that the control efforts may have reduced the TB prevalence; however, there were no data for a similar area not subjected to badger control, and no data for the concurrent prevalence in cattle. It was noted that removal of social groups of badgers infected with TB in areas of England and Wales had reduced the incidence of TB in cattle (Anderson, 1986). No details were given of the source of these data. Stuart & Wilesmith (1988) reported that in two areas of south-western England the incidence of TB in badgers dropped after intensive badger control, and in a separate study the incidence of TB in cattle dropped after badger control. The report did not state what happened in areas of no badger control.

Cheeseman *et al.* (1988) reported that transmission of TB from infected badgers to cattle may not always occur. In their study area in southern England, the last incident (till presumably mid 1986) of cattle infection occurred in 1980, yet TB was present in badgers in the area from 1981 to June 1986 with the prevalence ranging from 8% to zero. The latter absence of TB occurred on only three of 19 sampling occasions.

There are several other examples of field control of diseases of humans or livestock which appear to have had limited analysis of their effectiveness. Tuberculosis occurs in cattle and possums (*Trichosurus vulpecula*) in New Zealand. Coleman (1988) noted that control of possums reduced the incidence of TB in cattle though he did not give any supporting data. The use of fencing to control the spread of foot and mouth disease in parts in Africa (Taylor & Martin, 1987) was noted in section 3.7. As stated there, an analysis of the effect of such disease control efforts would be very interesting. Foot and mouth disease has been reported in wild boar (*Sus scrofa*) in the former U.S.S.R. (Kruglikov, Melnik & Nalivaiko, 1985; Boiko, Kruglikov & Shchenev, 1987) and control measures, such as hunting of boar, aimed at reducing the incidence of the disease in livestock have occurred. An evaluation of the effects of control would also be interesting. Note that in these situations the wildlife hosts may be native to the areas and apart from their disease status would not be classed as a vertebrate pest.

A negative result of attempted disease eradication is easier to interpret. Sheail (1991) described the efforts at rabbit and mosquito (myxomatosis vector) eradication and fencing (about 4.5 km erected in five days) in Kent in 1953 following the reporting of myxomatosis. The disease spread.

Field studies have also investigated the likely effects of rabies control. Trewhella *et al.* (1991) examined bait take by foxes in Bristol. If rabies entered Britain then rabies control may occur by poisoning of foxes so the data on bait take estimated what percentage of the fox population would eat a bait, and hence what the maximum percentage kill would be. The data would also provide a useful estimate of the maximum percentage of foxes that could be vaccinated if the bait contained a vaccine rather than a poison. The study involved marking bait with a compound that can later be identified in foxes. The number of foxes so marked compared to the number sampled is the estimate of the proportion that ate the bait, which in the study varied up to 35% of adult foxes. The estimation assumes that the foxes sampled were a random sample of the population, and that the eating of one bait was sufficient to kill or vaccinate a fox. The latter assumption is relaxed when the quantity of marker in the animal can be estimated and correlated with the amount of bait eaten. An evaluation of such a marker, iophenoxic acid, has reported a significant positive relationship between the amount of marker ingested by foxes and plasma-bound iodine concentration (Saunders, Harris & Eason, 1993). The relationship suggests that the number of baits eaten by a fox could be estimated, within a certain time period, by measurement of plasma-bound iodine concentration, though this was not tested empirically.

Hone & Pech (1990) estimated, by modelling, that a foot and mouth disease outbreak in feral pigs in Australia may cover 10 000–30 000 km^2 before first detection after about 30–50 days. In contrast, trials of control or eradication programmes for feral pigs have covered areas of only 50 km^2 (Hone, 1983), 120 km^2 (Saunders & Bryant, 1988), 74 km^2 (McIlroy *et al.*, 1989), 295 km^2 (Hone, 1990) and 94 km^2 (Saunders, Kay & Parker, 1990) so one hopes either that the trial results can be accurately extrapolated, that predictions are overestimates or that future trials will occur over bigger areas.

The economic analysis of control of infectious diseases is discussed in section 4.5 and modelling in sections 5.6 and 6.8.

3.17 Rodent damage control

Statistical analysis of rodent damage was reviewed in section 2.6.

3.17.1 *Problem definition and species*

The most common pest species of commensal rodents are black rat (*Rattus rattus*), Norway rat (*Rattus norvegicus*) and house mouse (*Mus domesticus* and *M. musculus*). A vast literature exists on rodent control, probably greater than for control of any other vertebrate pest. A complete review of that literature will not be attempted here.

3.17.2 *Methods and analysis*

Rodent control and its evaluation is dominated by poisoning, especially the use of anticoagulants. The literature on evaluation of rodent poisoning concentrates on the acceptability and toxicity of particular rodenticides to the target and non-target species, and on methods of detecting resistance to the rodenticides. Inferential and experimental methods have been used.

The statistical analyses used for evaluation of poisoning described in section 3.4 have nearly all been used for evaluation of rodent control. However, virtually all have concentrated on evaluation of toxicity rather than the evaluation of factors influencing toxicity. In contrast, Richards & Huson (1985) estimated a response surface for describing the relationship between percentage kill of rodents (the dependent variable) and the lethal feeding period of the poison that kills 98% of the rodents, the size of each bait placement and the frequency with which baits were replenished. The response surface accounted for 59% of the variation in percentage kill.

Inferential studies have been reported many times. Drummond (1985) reviewed evaluations of urban rodent control programmes in five locations – Hong Kong, Karachi, Amman, London and Kuwait. It was concluded that preliminary surveys of damage or abundance should precede control programmes and that such surveys need to occur subsequently so that effectiveness can be assessed. Only one study, in Amman, identified factors that affect the severity of a rodent problem, indicating that 79% of the variation in rat abundance was accounted for by the distribution of waste food, shelter and water.

The detection of rodenticide resistance in a field population is based on interpretation of trends, compared to a standardised regression, in rodent abundance over time since the start of poisoning (Drummond & Rennison, 1973; Rennison, 1977). The number of survey points where poisoned bait is taken relative to that at the start of poisoning (the dependent variable y), is graphed against the logarithm of days since the start of poisoning (the independent variable x). The plotted regression should occur between the 95% confidence limits of the standard regression. If the plotted proportion exceeds the upper standard limit then the rodents are resistant or the treatment is significantly less effective than a warfarin treatment if none of the rats were warfarin resistant. If no takes are recorded then it may indicate that all rodents have been killed or have stopped eating the bait. Quy, Shepherd & Inglis (1992) re-examined the analysis and concluded that as the method relies on bait consumption it may not be possible to establish why a poisoning fails to eradicate a rodent population. The failure may be caused by resistance, re-invasion or insufficient bait consumption.

The eradication of coypu from parts of Britain was assisted by intensive evaluation of the main method of eradication, trapping (Gosling & Baker, 1987; Baker & Clarke, 1988) and an evaluation of a trial eradication (Gosling *et al.*, 1988). The evaluation of trapping was described in section 3.6. In the trial eradication, the number of coypu trapped declined rapidly soon after the commencement of trapping then remained at low levels. The relative effects of trapping and cold winters on the annual change in coypu abundance were examined by multiple regression analysis (Gosling & Baker, 1989). Cold winters influence breeding success and juvenile survival. Trapping intensity had a highly significant effect on the change in coypu abundance and the severity of winters a significant effect. Together, the two factors accounted for 82% of annual variation in coypu abundance.

Movement of tagged coypu into part of a study area suggested that a trial eradication was not achieved there because of immigration. This was supported by analysis of the age structure of the population, which showed fewer coypu less than three months of age and more between six and 12 months of age than in a population outside the eradication area. The study compared trends in capture rates inside the study area with that of a similar marsh outside the area, illustrating the dramatic effect of trapping on coypu abundance. The results were used, with population models of coypu dynamics (Gosling & Baker, 1987), to recommend eradication of coypu from Britain. The model of coypu dynamics is briefly described in section 5.3.1.

The eradication of coypu has apparently been successful (Gosling, 1989; Gosling & Baker, 1989). The conclusion of eradication was based on the lack of trapped coypu and the absence of signs (faeces) within the area. The latter suggested that there were no trap-shy animals remaining, so increasing confidence in the conclusion of eradication.

Experimental studies by Buckle *et al.* (1984) described field tests of warfarin and brodifacoum for control of rice field rat (*Rattus argentiventer*) in a part of Malaysia. Two poisons and three baiting frequencies were evaluated in a randomised design with pre- and post-treatment assessment of rat populations and damage to rice stems. The experiments used experimental control areas, with no poison, and the replication was sequential. The results were analysed by analysis of variance and showed significant differences between baiting frequencies and replicates, and a significant interaction of poison type and replicate. The latter was associated with a decline in the effectiveness of warfarin in later replicates, but no apparent decline in the effectiveness of brodifacoum.

The effects of poisoning on mouse damage to soybeans in western New South Wales was estimated by a split-plot experimental design and analysis of variance (Singleton *et al.*, 1991a). Poisoning had no significant effect except at the level of sites within transects wherein there was a significant interaction of poisoning, farms and distance.

The economic analysis of rodent damage control is discussed in section 4.6 and modelling in sections 5.7 and 6.9.

3.18 Control of bird strikes on aircraft

The statistical analysis of problems caused by birds striking aircraft were described in section 2.7. The issue here is the control of such bird strikes.

3.18.1 *Problem definition and species*

Theoretically, the number of bird strikes could be reduced by reducing the number of birds, or changing the flight times of the aircraft or of the birds, and the effects of bird strikes reduced by strengthening the aircraft. In practice, most efforts have gone into reducing the number of birds. This is understandable, for example, an aircraft cannot be strengthened to withstand collision with a swan and expect to be a light and cheap aircraft for commercial operations. Wright (1968) and Brough & Bridgman (1980) assumed that a reduction in the number of birds will lead to a decrease in the number of bird strikes. The assumption appears to be untested experimentally.

3.18.2 *Methods and analysis*

The topic of bird strike control has been written about in many publications. However, there are few scientific studies where methods of control of bird strikes have been evaluated. On closer inspection this could be for good reasons. There are very limited opportunities for studies with replicated or experimental control sites as there is commonly one airport being studied. The airport typically cannot be moved and the flights cannot be stopped or flight times changed. Inferential and experimental studies have been reported.

An inferential study of bird strikes at Sydney airport reported that the number of strikes decreased after 1964 when nearby garbage was removed, airport drainage was improved, the frequency of grass mowing was increased and airfield margins were sprayed with insecticide (van Tets, 1969). The relative effects of each action could not, however, be estimated. In fact, the change in the number of bird strikes was not statistically tested so the reduction may have been more apparent than real.

The curvilinear relationship between abundance of birds and frequency of bird strikes reported by van Tets (1969) implied that a substantial reduction in bird abundance was necessary to reduce the number of bird strikes. Van Tets (1969) cited a reduction of 80% in bird abundance being needed. My analysis of that data in Chapter 2 suggests that the effect of reduction of bird abundance is more sensitive because of the linearity of the relationship. However, the lack of significance of the relationship reported by Burger (1985) suggests that further work is needed to understand the expected response in bird strikes to changes in abundance of birds.

An experimental study by Brough & Bridgman (1980) reported the effects on bird abundance of the length of grass at British airports. The grass for about 90 m alongside the runway at each airport was maintained at a height of 15–20 cm, except where siting of equipment prevented this, in which case grass was mown to a height of about 5 cm. Elsewhere on each airport the grass was mown to a height of about 5 cm. The growing of long grass had apparently been recommended in 1949 but this study was the first to evaluate the recommendation.

The effects of grass height were estimated by three-way analysis of variance. The three factors were grass height, airport and bird group. Data from two studies were described. The mean number of birds was about five times higher on the areas of short grass than of long grass, a difference that was highly significant. The effects of grass height differed significantly between groups of birds, and between airports. The greatest effect was shown by gulls and lesser effects by lapwings (*Vanellus vanellus*) and starlings (*Sturnus vulgaris*). In the first study, birds were less abundant in the long grass at nine of 10 airports but at the tenth airport the reverse occurred. The latter result was attributed to sparse vegetation, so that even when grass was high it was not thick.

The study could not assess the effects of grass height on the number of bird strikes as both grass heights occurred at each airport. Brough & Bridgman (1980) discussed this and other limitations to the evaluation of control methods at airports. In particular, the bird strike rate is usually very low so it may take data from many airports or many years to accumulate enough data for analysis. Over such large numbers of airports or years there will be many differences or changes that may mask any effects of the control methods.

The economic analysis of control of bird strikes on aircraft is briefly discussed in section 4.7. Mathematical modelling of bird strikes is discussed in sections 5.8 and 6.10.

3.19 Control of bird damage to crops

The topic of bird pests and crops was introduced in section 2.8. The analysis of the effects of bird control is the emphasis here.

3.19.1 *Problem definition*

Many methods of bird control are used, such as scaring, trapping, poisoning, shooting, spraying, explosives and netting (Bruggers, 1989; De Grazio, 1989; Meinzingen *et al.*, 1989).

3.19.2 *Methods and analysis*

The methods of study are laboratory and field based and analyse effects of control on either bird abundance or bird damage. Occasionally both are analysed.

Lacombe, Matton & Cyr (1987) studied the effect of a chemosterilant on aspects of reproduction in captive red-winged blackbirds. Two populations of blackbirds were treated, each group for a different duration, and the results compared, by analysis of variance or a Kruskall–Wallis test, with results from an experimental control group which had not been treated. Testicular weight did not differ between the treated and control groups, but the number of spermatocytes and other responses did differ significantly.

Dolbeer (1988) reviewed lethal methods of reducing bird damage in agriculture. The review was broader than the one topic of bird damage to crops and concluded that in many studies the effects of lethal control on bird abundance had been estimated but the effects on bird damage had not. Many studies also estimated the effects on abundance at one site without study of bird abundance in at least one area that had no bird control. Feare, Greig-Smith & Inglis (1988) reviewed non-lethal methods of reducing bird damage. That review was more optimistic for successful control though that may not necessarily reflect a higher standard of evaluations.

Murton & Jones (1973) tested the effect of shooting wood-pigeons (*Columba palumbus*) on damage to cabbages and sprouts in a part of England. In each of two years a shooting site and an experimental control (no shooting) site were studied and indices of damage recorded. Hence, the second site was a simultaneous experimental control. There were no significant differences in damage between the sites in either year, as assessed by a Mann–Whitney U-test. In a subsequent year when shooting did not occur damage at both sites was less than in the previous two years. The efficiency of shooting methods differed (Murton & Jones, 1973). The method of roving round fields resulted in fewer birds being killed and corresponded to the greater reduction of damage. That is, it was better to have a person walking round a field scaring the birds than in a hide shooting them.

Another study of control of wood-pigeons in England reported that shooting of the birds around roosts killed large numbers of birds but that it did not increase winter mortality above that in the absence of shooting (Murton, Westwood & Isaacson, 1974). When shooting ceased, there was no increase in abundance of wood-pigeons. An alternative shooting method, that of using a decoy to attract wood-pigeons, was evaluated. The percentage of a flock that responded to the decoys declined as flock size increased, as assessed by regression analysis. The maximum number or percentage of a flock that were shot occurred at intermediate numbers of decoys; fewer or more decoys resulted in fewer birds being shot. My analysis of data in their Table 2 shows that at one site, Carlton, the number of pigeons shot was highly correlated ($r=0.92$, df$=12$, $P<0.01$) with the hours expended in shooting.

A study of an acoustic device to reduce damage to grapes by grey-breasted white-eyes (*Zosterops lateralis*) in south-western Australia used simultaneous experimental control plots nearby plots which had the device (Knight & Robinson, 1978). The number of damaged grapes exceeded 3000 in each of the experimental control plots and was less than 500 in each of the four treated plots. There was no statistical analysis of the data. The study was replicated in two other trials though no details of the results were reported.

The economic analysis of control of bird damage to crops is discussed in section 4.8 and modelling in sections 5.9 and 6.11.

3.20 Rabbit damage control

The statistical analysis of rabbit damage was reviewed in section 2.9. The analysis of control of such damage is the topic here.

3.20.1 *Problem definition and species*

The agricultural and ecological effects of rabbits (*Oryctolagus cuniculus*) were reviewed by Sumption & Flowerdew (1985). They described the results of studies in Britain, continental Europe and Australia on changes in vegetation and wildlife after introduction of myxomatosis. The effects of rabbits were studied mostly during and after control rather than before and after, and hence have less explanatory ability.

The analysis of rabbit control has concentrated on studies of control versus no control, or types of control. This has helped the development in Australia of many states using one poison, sodium monofluoroacetate, at different concentrations in baits, despite the apparent absence of studies to show which concentration is most effective.

3.20.2 *Methods and analysis*

The analysis of the effects of myxomatosis in the field has been mostly inferential. Myxoma virus was introduced into Australia in the 1940s and 1950s to control rabbits. Details of the analysis of its effects were described in section 3.11 on biological control. These data refer to myxomatosis on the Australian mainland. An evaluation of myxomatosis for rabbit control on the subantarctic Macquarie Island reported variable results (Brothers *et al.*, 1982). In some areas there were rapid declines in rabbit abundance, as assessed by standardised monthly counts, but in other areas of the island there was very little effect. It was estimated that overall myxomatosis introduction from 1978 to 1980, inclusive, had depressed rabbit abundance by about 60%. The disease was introduced via infected vectors, fleas. The disease was observed to spread about 3 km over 12 months, though the fastest rate was 2.2 km in 44 days. The assessment of the effects of myxomatosis was mainly by use of a sequential experimental control, that is, counts before and after disease introduction.

The effects of ripping warrens on rabbit abundance in a part of semi-arid New South Wales was reported in an experimental study by Parker, Myers & Caskey (1976). Areas with and without ripping were measured with counts of warrens and active entrances used as indices of rabbit abundance. The study occurred from 1965 to 1974 inclusive, which included a drought and period of above average rainfall. The data on rabbit warrens were not statistically analysed. The interpretation of the result was complicated, as described by Parker *et al.* (1976), because the actual areas within which ripping occurred changed between years, as did the areas with no ripping. Hence, it would be expected that as the area ripped increased then the remaining area (unripped) became smaller and contained fewer warrens and even fewer warrens per unit area, as the farmer would rip warrens not areas with no or few warrens. An extra experimental area, that did not move over years, was used on adjacent properties. There the number of warrens fluctuated with rainfall, but generally increased.

The interpretation of such field data would be aided by not repeatedly shifting the goal posts. If, however, after the start of the experiment areas to be treated and not treated were selected at random, then the different areas would be experimental replicates amenable to analysis of variance. Such analysis would require estimates of pest abundance before and after ripping.

Cooke (1981) and Cooke & Hunt (1987) used factorial experimental designs to test the effects of different control methods on rabbit abundance. Both studies used replicated sites for each treatment (control method) and experimental controls (no control) but Cooke & Hunt (1987) used blocks as sites for replication, allowing variation associated with blocks to be removed in the analysis of variance and hence giving a more precise analysis of the effects of treatments.

The first study estimated the effects on rabbit abundance of poisoning, ripping and fumigation singularly and in all combinations. All three methods reduced rabbit abundance and the combination of two or three methods reduced rabbits to the lowest abundance. The second study estimated the effects on rabbit abundance of ripping, poisoning and poisoning followed by ripping. All treatments reduced rabbit abundance with the greatest reduction occurring after ripping and poisoning followed by ripping. In both studies it was argued that the effects on rabbit abundance of treatments, such as poisoning and ripping, were multiplicative, so for analysis of variance, data were transformed to logarithms to give additive effects. Additivity of treatments is an assumption of analysis of variance (Snedecor & Cochran, 1967).

Foran *et al.* (1985) tested the effects on rabbit abundance and vegetation of the control treatments, ripping with fumigating, 1080 poisoning, and 1080 poisoning with ripping and fumigating. Experimental controls were included in the randomised experimental design. Analysis of covariance, which adjusted rabbit abundance for pre-treatment levels, showed significant effects of control on rabbit abundance, with the lowest abundance after ripping and fumigation. Rabbit control did not have a significant effect on the frequency of occurrence of any plant species, or on standing biomass of the vegetation. It was concluded that rainfall had a greater effect on the vegetation than rabbit control during the study. Foran (1986) reached similar conclusions after multivariate analyses of classification and ordination to determine the effects of rabbits and rabbit control.

The economic analysis of rabbit damage is discussed in section 4.9 and modelling in sections 5.10 and 6.12.

3.21 Control of predation of rock-wallabies by foxes

3.21.1 *Problem definition and species*

The dynamics of remnant populations of rock-wallabies (*Petrogale lateralis*) and the associated effects of fox (*Vulpes vulpes*) control were studied experimentally by Kinnear *et al.* (1988). The rock-wallabies were confined to rocky outcrops in south-western Australia and shared the outcrops with foxes, which were recent colonists. The impetus for the study was the very low numbers in several rock-wallaby colonies and the presence of foxes. Predation on rock-wallabies by foxes was not reported as having been observed before or during the study. Hence, the evidence was circumstantial, as noted in the paper. Study of such predation would be worthwhile, and could use some of the methods described in section 2.4 to study predation of livestock.

3.21.2 *Methods and analysis*

Two populations of wallabies occupying rock sites not subjected to fox control declined by 14% and 85% and a third population increased by 29%. Fox control was attempted by baiting with sodium monofluoroacetate (compound 1080) and two populations in areas where fox control occurred increased by 138% and 223%. It was concluded that fox predation was probably an important factor in the decline of the wallabies.

The evidence presented seems convincing but closer examination of the data is warranted. Surprisingly, the statistical significance of the effect of fox control on wallaby abundance or rate of increase was not reported. The data presented by Kinnear *et al.* (1988) provide an opportunity to test for such effects and re-examine the conclusions of the study. The tests are now described.

The effects of fox control on wallaby abundance could be tested by analysis of variance. Fox control would be a factor, and years a second factor, in a two-way analysis. The effects of fox control are assumed to be fixed and years could be tested as fixed or random factors depending on the assumptions in the study, but are assumed here to be random as there was apparently nothing special about those particular years. Fox control was replicated but the experimental control (no fox control) was not replicated with a site in which data were collected in each of the same years. The data for analysis is a sub-set, because of missing data, of that originally presented and was for two sites with fox control (Nangeen and Mt. Caroline) and one site with no fox control (Sales Rock) for the years 1979, 1982, 1984 and 1986 (Fig. 3.5).

Analysis of variance of the data, after transformation to common logarithms to improve variance homogeneity, found no significant ($P > 0.05$) effects of fox control, years or the interaction of fox control and years (Table 3.4a). This suggests that fox control had no effect on wallaby abundance.

A better analysis of such data was described by Huson (1980). The data could be analysed as the abundance of wallabies at each site relative to the initial abundance. Hence data for a further analysis were the abundance

Fig. 3.5. Trends in numbers of rock-wallabies at three sites in south-western Australia. At two sites (Nangeen ---- and Mt. Caroline ········) fox control occurred starting in mid-1982 and at one site (Sales Rock ——) no fox control occurred. Data are from Kinnear *et al.* (1988).

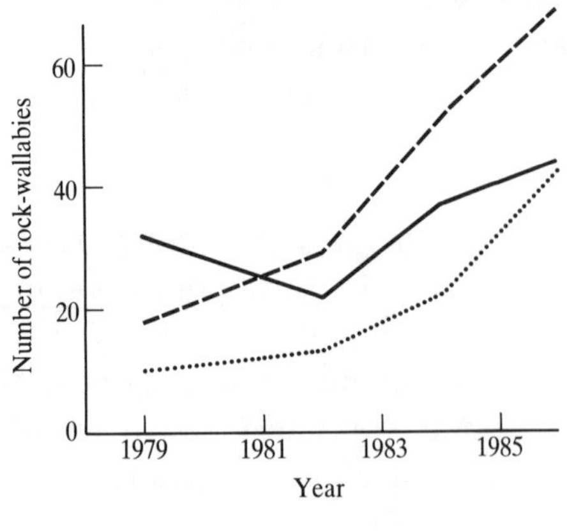

expressed as a percentage of the initial population size. Such an analysis is method 6 in Table 3.1. Each population was assumed to start at 100%, though the data for the first year (1979) was not included in the analysis. There was a significant effect of years, but no significant effect of fox control or the interaction of control and years (Table 3.4b).

The data can also be used to estimate the observed instantaneous rate of increase of abundance for each rock-wallaby population. As described by Caughley (1980) the rate is estimated by the regression of abundance, transformed to natural logarithms, on time. By hypothesis, the rate of increase should be zero for stable populations. Calculations show that the observed rate of increase of each wallaby population at the two sites with fox control (0.20/year) was significantly ($P<0.05$, df$=2$) different from zero, with the correlation coefficients (r) being 0.993 and 0.961. In contrast, at the site where there was no fox control the rate of increase of the wallaby population was

Table 3.4. *Analysis of variance of the effects of fox control, years and their interaction on the abundance of rock-wallabies*

(a) Control and years are analysed as fixed and random factors respectively. Data were transformed to common logarithms for analysis. All F ratios were not significant.

Treatment	df	Mean squares	F ratio
Fox control	1	0.025	0.926 (=0.025/0.027)
Years	3	0.130	2.766 (=0.130/0.047)
Control × years	3	0.027	0.575 (=0.027/0.047)
Residual	4	0.047	
Total	11	0.681	

(b) Data were expressed as the percentage of the initial population size at each site.

Treatment	df	Mean squares	F ratio	Significance
Fox control	1	50244.500	8.247 (=50244.500/6092.667)	NS
Years	2	28240.333	26.900 (=28240.333/1049.833)	*
Control × years	2	6092.667	5.803 (=6092.667/1049.833)	NS
Residual	3	1049.833		
Total	8	15257.500		

* $P<0.05$, NS = not significant.

0.06/year which was not significantly different ($r=0.56$, df$=2$, $P>0.05$) from zero. The analysis of rates of increase is essentially an analysis of the interaction of years and fox control. The results of this analysis support the conclusions reported by Kinnear *et al.* (1988) that foxes have an important effect on rock-wallaby population dynamics.

The case study highlights the need for detailed analysis of data before conclusions are reached. Also, the conclusions are sensitive to the type of analysis. This was probably accentuated in the study by the small data set and an expected lag time in a detectable response by the rock-wallaby populations to the reduction in predation by foxes. Ideally, data in the original study should have been statistically analysed. A similar conclusion is needed for the study of fox predation of numbats (*Myrmecobius fasciatus*) in south-western Australia (Friend, 1990). The conclusion that poisoning of foxes resulted in a more rapid growth of the numbat population was not based on a statistical analysis of abundance or rate of increase.

3.22 Non-target effects of control

3.22.1 *Problem definition and species*

Control methods such as, for example, poisoning and trapping, may catch, harm or kill species other than the target vertebrate pest. Such may be of conservation or animal welfare concern. The effects on non-target species can be studied by the same methods used to assess the effects of control on the target species.

There may be some differences, however, with the estimation of effects on non-target species. The non-target species may be rare or endangered, so effects on the populations may be of conservation concern. For such species, the effects of pest control on individuals may be as important as the effects on the population. This is in contrast to the target species where most effort is concerned with the population rather than the fate of individuals.

A non-target species may be particularly susceptible to a pesticide, which may result in a high percentage kill of its population. Control of the target pest species may reduce its abundance and allow a non-target species to replace it as a pest (Walker & Norton, 1982). This may result from release from previous competition, a switch in the diet of the non-target species or changes in predator–prey relationships. Lastly, the non-target species may be rare or difficult to census so the effects of control will be difficult to estimate.

3.22.2 *Methods and analysis*

Six methods have estimated the effects of control on non-target species. Laboratory tests of sub-lethal effects (method 1) by King, Oliver &

Mead (1978) and Twigg & King (1991), and related papers, used the levels of plasma citrate as an index of tolerance of sodium monofluoroacetate (compound 1080) in many species. Citrate is formed from 1080 within the body and circulates in the blood (Mead *et al.*, 1985). McIlroy (1981, 1986), and related papers, described the results of laboratory toxicity trials (method 2) with many non-target species of wildlife in Australia.

O'Brien *et al.* (1986) compared the estimated LD_{50} for non-target species with the amount of poison, sodium monofluoroacetate, in vomitus of penned feral pigs after the pigs had eaten poison (method 3). The level of poison was higher than the LD_{50} for 14 of 41 species. Hence the vomitus, if eaten, posed some hazard to those non-target species.

Field tests of toxicity are method 4 for evaluating effects of pest control on non-target species. McIlroy *et al.* (1986b) reported the evaluation of a field test of the toxicity of 1080 in meat baits to non-target species during control of wild dogs (*Canis familiaris*) in south-eastern Australia. The baits were placed on the ground along roads and tracks. The effects of birds were assessed by bird censuses before and after poisoning, and tested by Student's *t* test. The effects on small mammals were assessed by trapping before and after poisoning and analysed by Chi-square tests. There was a significant decline in abundance of one bird species, spotted pardalote (*Pardalotus punctatus*), which was probably a spurious result, as noted by McIlroy *et al.* (1986b), as the bird forages on insects in the tree tops and very rarely on the ground where the bait had been placed. There were no significant effects on small mammals.

Simpler methods of estimating the lethal effects on vertebrate pest control (method 4) have been reported. The simplest is careful and systematic searching for and tallying the number of dead non-target animals after pest control. Hone & Pederson (1980) noted the number of bird species found dead after poisoning of feral pigs in a part of southern Australia. Newsome *et al.* (1983) reported the number of non-target species caught in traps set for dingoes (*Canis familiaris*) in south-eastern Australia. The number of non-target species per trap night was significantly different, as assessed by Chi-square analysis, between types of trap. Bjorge & Gunson (1985) tallied non-target deaths after wolf control in Alberta. Meinzingen *et al.* (1989) reviewed several studies of quelea control in Africa and noted the numbers of non-target birds killed after spraying of quelea. The effects on the populations of the non-target species were not described in any report. Such tallies should be supplemented by chemical analysis of carcasses to confirm that deaths were associated with pesticide contamination. The absence of any known deaths of non-target species may be ambiguous; either none died or some died but were not found.

Yom-Tov (1980) concluded that a control poisoning of chukar (*Alectoris chukar*) in Israel resulted in no apparent deaths of raptors. This was based on an absence of raptor carcasses but no data were reported on abundance of the raptors before or after the poisoning, or in an area where poisoning did not occur.

The detection of sub-lethal effects of vertebrate pest control (method 5) has been used in several places. Saunders & Cooper (1982) reported in an area of southern Australia that pesticide residues were common in birds of prey at the same time as pesticide control of house mouse was occurring. The significance of the residues was not assessed. The presence or absence of pesticide residues in barn owls (*Tyto alba*) in parts of New Jersey was studied by Hegdal & Blaskiewicz (1984). The pesticide, brodifacoum, was being used for control of rats and mice. Owls that had been found dead, because of accidents or other causes, were sampled for residues. The results must be interpreted with caution as when few individuals in a population have residues there is a low probability of detecting those individuals. These topics of surveillance and sample sizes needed to detect phenomena such as residues and diseases were discussed in more detail in section 2.5. Surveys of residues in living birds would also assist in estimating non-target effects of poisoning.

Brunner & Coman (1983) and Bryant, Hone & Nicholls (1984) studied bait acceptance by non-target birds (method 5) in southern Australia. They offered coloured grain and compared intake with that of natural (uncoloured) grain offered simultaneously. The grain is used for control of rabbits or feral pigs. Intake of coloured grain, especially green and blue, was significantly less than that of uncoloured grain, suggesting that bird intake of grain poisoned for mammalian pests may be reduced by such colouring. Brunner & Coman (1983) analysed data by Student's *t* tests and Bryant *et al.* (1984) by split-plot analysis of variance. Hone *et al.* (1985) reported that, at a field site in western New South Wales, neither feral pigs nor free-flying birds had differences, on average, in intake of dyed or undyed grain over a two-year study. However, intake differences were highly variable as a result of significant interactions of years, seasons, grains and times within seasons as assessed by analysis of variance. This study, an example of method 5, tested the effects of years, seasons, times within seasons, grains and dye colours as multiple factors in a factorial experimental design. The results suggest dyeing grain may have negligible value for reducing intake by birds in the field if the grain is available to the birds for periods up to two weeks.

The likely effects on non-target species can be studied by modelling (method 6). Some methods and results of such studies are described in section 6.13.

3.23 Conclusion

Many of the comments made in the conclusion of Chapter 2 apply equally in this chapter. The literature reviewed shows a mixture of elegant experimental design through to naturalistic or inferential studies. There is scope for greater use of statistical analyses, testing of alternative hypotheses and more replication. Many studies used inadequate replication and randomisation, and needed more thorough comparisons of observed with expected results. The latter can be estimated by modelling as described in Chapters 5 and 6. A summary of some of the worse and better features of the analyses is given in Table 3.5.

In the evaluations in this chapter, a series of methods were described ranging from laboratory to field tests. In each there is a potential problem in

Table 3.5. *Comparisons of aspects of experimental designs for evaluation of vertebrate pest control. A better feature is defined as one that gives more sensitive estimates of treatment effects because of the reduction in experimental error (residual variance)*

Worse feature		Better feature
No experimental control	⟶	At least one experimental control
No replication	⟶	Replication
Sequential or simultaneous experimental controls	⟶	Sequential and simultaneous experimental control
Simple randomised design	⟶	Design with sources of variation controlled such as randomised blocks
Design with no pest control versus pest control	⟶	Design with no pest control versus many levels of pest control

Table 3.6. *Differences in the accuracy, precision and relevance of methods of study of damage and control of vertebrate pests*

Research method	Accuracy	Precision	Relevance
Laboratory	Higher	Higher	Lower
Pen	↕	↕	↕
Small plots			
Field	Lower	Lower	Higher

Table 3.7. *Checklist of steps in estimating and analysing damage and response of pests or damage to pest control. The list is relevant for studies in the laboratory, pen, field or modelling. Topics with a question mark are to be avoided or minimised*

1. Experimental design
   - scheme including control
   - randomisation
   - replication
   - local control for similar initial conditions
   - fixed or random effects
   - covariates
   - confounding?
   - pseudoreplication?
   - effects of experimentation?
   - pseudodesign?
2. Sampling
   - scheme
   - randomisation
   - replication
   - sample size
   - replacement procedure
   - precision required
   - pseudoreplication?
3. Analysis
   - test of difference or relationship (association, goodness of fit, relationship)
   - specific test
   - mean, variation and frequency distribution of variable
   - level of significance
   - validity of assumptions and robustness to deviations from assumptions
   - likelihood of type I or type II errors
   - comparison of observed and expected (modelling) results
4. Extrapolations
   - laboratory to field
   - small plots to larger areas
   - one population to many
   - small to large population
   - closed to open population
   - non-lethal component of control to lethal control
   - model to laboratory, pen or field
5. Identify spatial and temporal sources of variation in response
   - pest abundance
   - environmental factors
6. Comparative costs of alternative designs, sampling and analyses

extrapolating from the more controlled environment of the laboratory to the more variable field situation (Table 3.6). Dillon (1977) described an analogous problem in agricultural research as did Hairston (1989) in ecological research. Dillon (1977) reported a relationship between the ratio of crop yield on farms to crop yield in experiments and the average area of each crop. As crop area increased the ratio dropped, indicating the high yields of experiments were not repeated in large crops on farms. Similarly, in control of vertebrate pests the large effect of control methods reported in laboratory or semi-field experiments may not be repeated in extensive field use.

The analysis of response to control of pest abundance or damage involves several steps. These are summarised in Table 3.7, which should be used to complement Tables 2.2 and 2.6. The wily trapper described at the start of this chapter could use the checklist. More importantly from the perspective of this book, scientists observing the trapper should use the checklist.

The statistical analysis of the effects of pest control on damage or pest abundance has been described. A set of analyses for estimating the economic value of pest control are available and are described in the next chapter.

4

Economic analysis

Money may make the world go around but it has not been the driving force behind analysis of vertebrate pest control. Surprisingly few efforts at vertebrate pest control have involved even a cursory economic analysis. Hence, this chapter contains more theory than empirical examples. The outcomes of decisions about when, where and how to control pests or their impacts must involve use of resources and so must have costs and hopefully some benefits.

An economic evaluation of the damage by or control of vertebrate pests is essential to an understanding of the pests' role in a production or conservation system. Cherrett *et al.* (1971) considered that the assessment of the economic status of a pest should be simple. The expected damage and costs of counter measures would be assessed. If the damage cost was greater then control was economically worthwhile. An interesting example of the link between economics and ecology was described by Sukumar (1991) for elephants (*Elephas maximus*) in southern India. Adult male elephants comprised about 7% of the population but caused 62% of the economic loss to crops. The difference was associated with their greater size and differences in behaviour.

Dillon (1977) discussed the difference between statistical and economic significance in experiments. Scientific research applies tests of significance to data ('the cult of the asterisks'). Statistical significance at a $P<0.05$ level may not be the best criterion for economic evaluation. For example, a farmer is more interested in how much pest control to use and the economics of control than whether there is a 5% chance or less that a response to control exists. The number of detailed economic evaluations of vertebrate pest control is less than the number of detailed statistical evaluations. Dyer & Ward (1977), Caughley (1980), Dahlsten (1986) and Dolbeer (1988) were critical of the lack of economic evaluation of vertebrate pest control.

This chapter examines methods for such economic evaluation. The coverage of economic topics is not exhaustive. It is intended as a guide with details of more comprehensive analyses in the references provided.

4.1 Objectives

The objectives of an economic evaluation can be varied. They usually reduce to one or more of the following: to maximise profit, to estimate cost-effective levels of control inputs, or to satisfy legislation. A distinction is made between profit (the monetary difference between returns and costs) and benefits (monetary return plus non-monetary returns). In an eradication programme the objective may be to determine the least cost or the most cost-effective actions. An alternative objective may be to estimate the threshold pest density above which the benefits of control will exceed the costs of control (Norton, 1976). Birley (1979) described two alternative objectives of pest control: maximise profit or minimise the combined costs of damage and control. The specific analysis used may vary with the objective of management and hence a variety of analyses are now examined.

Chiang (1979) described an economic threshold of a pest population to be the population size at which control should be initiated. In contrast, Birley (1979) described the economic threshold as the level of pest abundance at which the cost of control is equal to the cost of damage prevented by such control. A different definition of a threshold was used by Begon *et al.* (1990). They described an objective of pest control as the reduction of a pest population to a level at which no further reduction is profitable. This level of abundance, or population density, was called the economic injury level (EIL). It occurs when the extra (marginal) costs of control equal the extra (marginal) benefits, which are described further in section 4.2.1. Begon *et al.* (1990) described two limitations to the concept. Firstly, it implies that the effects of pest control are immediate, when there will usually be a lag time. Secondly, the EIL may vary over time with commodity prices.

4.2 Types of analysis

Four analyses are described here: marginal cost–benefit, cost–benefit, cost-effectiveness and decision theory. The types of economic analysis used are described in many textbooks. The general field of biological resource economics is discussed by McAllister (1980) and Clark (1976, 1981), and the economics of pest control by Conway (1981) and Tisdell (1982).

4.2.1 *Marginal cost–benefit analysis*

Marginal cost–benefit analysis is concerned with estimating the level of inputs at which the marginal cost of the extra unit of input equals the marginal benefit of that unit. At this level of input the activity makes maximum profit (Fig. 4.1). In our situation inputs are control efforts. Conway (1981) showed a similar graph to that of Fig. 4.1, though marginal returns appear to be graphed incorrectly by Conway.

Begon *et al.* (1990) illustrated the estimation of the pest density at which maximum profit occurs. They explained the estimation of the cost of reducing pest density to particular levels. It costs little to reduce pest density by a small amount when starting from a high density. Each incremental reduction in density costs slightly more so the cost rises exponentially (Fig. 4.2). Simultaneously, economic returns decline as pest density increases. The pest density at which maximum profit occurs is when the maximum difference exists between the returns and the costs (Fig. 4.2). That level of pest density is called the economic injury level (EIL). If the pest population density is reduced below the EIL then the difference between returns and costs will be less than at the EIL.

Marginal analysis assumes that there are smooth, continuous relationships between the control inputs and costs or returns. The estimation of such relationships, especially the returns function, requires careful use of good experimental design (Dillon, 1977). The estimation is by regression analysis, which is typically linear for the total cost function and non-linear for the return function. As discussed by Dillon (1977) response relationships are more efficiently estimated by designing studies with increased levels of the control inputs and with consequent fewer replications for any particular level (see Fig. 2.3).

Tisdell (1982) used a form of marginal analysis to examine theoretical aspects of managing wild pigs (*Sus scrofa*) and their impacts relative to their commercial value and control costs. It was concluded from an analysis of a

Fig. 4.1. Expected relationships between (*a*) total costs and returns and the levels of control inputs, such as amount of pesticide, number of traps, and (*b*) marginal costs and returns and levels of control inputs. Total costs equal total returns at *f*, and maximum profit occurs at *c*, the level of inputs at which the marginal cost equals the marginal return.

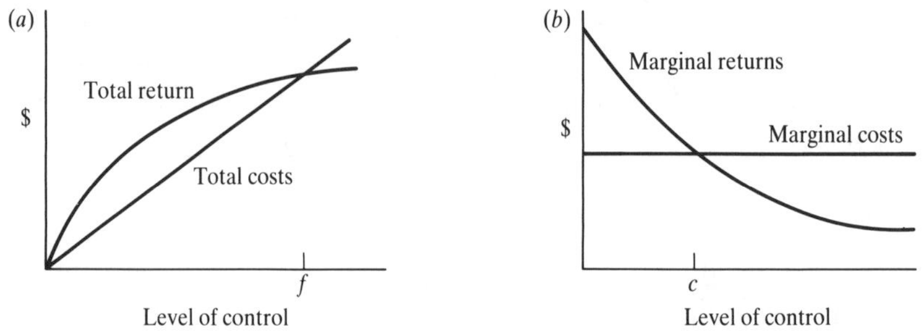

model of pest impacts (Headley, 1972a) that (i) maximum profit may be above a threshold of abundance for pest damage, (ii) greater control could be expected when there are high prices for the product being damaged and conversely lower levels of control when prices are low, (iii) the greater the damage by each individual pest (the marginal loss) then the greater is the profit incentive for pest control, and (iv) conversely when the marginal cost of reducing pest abundance is high then the smaller is the expected reduction in pest abundance.

Izac & O'Brien (1991) assumed that the total costs and benefits of control increased as a feral pig population was reduced in abundance. Four scenarios were described that resulted in different levels and outcomes of control: benefits always greater than costs (produced eradication), costs always greater than benefits (no control), benefits greater than costs when reduction is large (eradication), and costs greater than benefits when reduction is large (intermediate control).

Headley (1972b) described a simplistic model of production, for example of a crop, and the effects of pests and pest control. The model assumed a curvilinear relationship between the level of inputs and level of outputs (Fig. 4.3). As inputs increased then output increased but at a decreasing rate (diminishing returns). When pests occurred in the system levels of output were lower. If pest control occurred then production was higher than that for no control with the difference related obviously to how effective the control had been (Fig. 4.3). The model has not been tested empirically.

4.2.2 *Cost–benefit analysis*

Cost–benefit analysis (CBA) is a commonly used technique that estimates the monetary value of benefits and costs of a particular level of

Fig. 4.2. Hypothetical relationships between costs and returns of an economic activity and pest population density. The arrow shows the pest population density at which maximum profit occurs – the economic injury level. (After Begon *et al.*, 1990.)

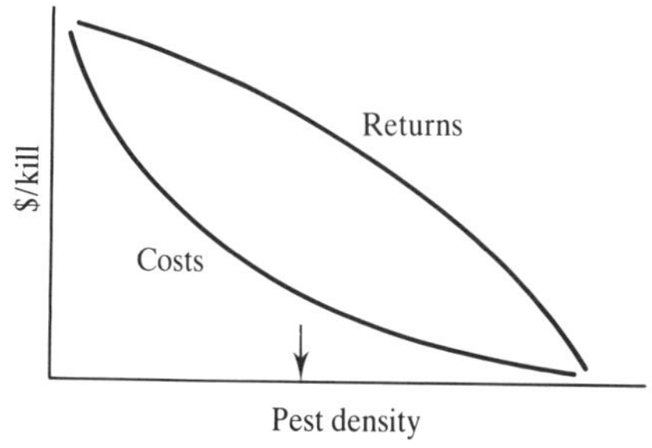

activity. If the benefits exceed the costs then the ratio of benefits/costs is greater than one and the proposal will be economically profitable. To estimate the level of activity that will maximise profit requires estimating many such ratios. Cherrett *et al.* (1971) noted that though the ratio of benefits to costs was a good criterion for pest control, the ratio was often not estimated because of the practical difficulties of doing so. Estimates were best when rates of pest infestation and monetary return were steady for some time.

Benefits are the gain in something desirable, and the reduction in something not desired. Costs are the lost opportunity of a benefit, and the increase in something not desired (McAllister, 1980). The costs may be private or social, in the latter case being broader and including costs being borne by those who do not stand to benefit but who may suffer. Tisdell & Auld (1988) discuss the difference between private and social cost–benefit analysis, but relative to biological control of weeds.

The advantages and disadvantages of CBA are discussed by McAllister (1980) and in a refreshingly entertaining manner by Sinden (1980). Both include detailed discussions of the advantages and disadvantages of the analysis, and in Sinden's case, a useful critical review of selected texts and papers to that date. He concludes that the literature on the analysis has increased greatly without a similar increase in its capacity to solve problems.

An example of the use of CBA is to differentiate between alternative possible actions. When the costs and benefits can be estimated, alternative criteria would result in differing actions being selected (Table 4.1). The action that

Fig. 4.3. Hypothetical relationships between inputs, such as the amount of fertiliser added to a crop, and outputs, such as yield, of a production system and the effects of pests and pest control on the relationship. (After Headley, 1972b.)

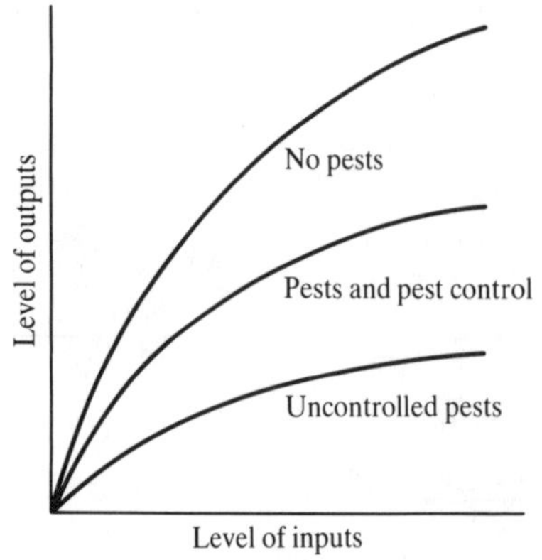

minimises costs is action 3, that which maximises benefits is action 1, that which maximises the difference between benefits and costs is action 1, and that which maximises the benefit/cost ratio is action 3.

The benefit/cost ratio of deer control in cornfields was estimated by Hygnstrom & Craven (1988) using the formula:

$$\text{benefits/costs} = (\text{yield}_c - \text{yield}_t)/(\text{materials} + \text{labour costs}) \quad (4.1)$$

where yield_c was the average yield lost on the experimental control (no deer control) and yield_t was the average yield lost on the treatment fields (deer control). The yield loss data appear, however, to have been in physical units (kg/ha), or the conversion to dollars was not shown, and the cost data in dollars. Both sets of data should be in dollars.

Tisdell, Auld & Menz (1984a) reviewed the need for cost–benefit analysis in assessing biological control of weeds. They also stressed the need for dissemination of control information and investment in research by governments, rather than private industry and individuals.

Norton (1976) described a method for estimating profitability in pest control that has elements of marginal analysis and cost–benefit analysis. The reduction in damage as a result of control is estimated in monetary terms and compared with the cost of control. If the reduction is greater than the control cost then it is profitable to carry out control. The reduction in damage is estimated as:

$$\text{Reduction in damage} = p\ d\ l\ k \quad (4.2)$$

where p = price per unit of product (for example, \$/tonne), d = the damage coefficient, expressed as for example tonnes/ha for each individual pest, l = the level of pest attack, for example the number of pests, and k = the proportional reduction in damage associated with the control method. If the cost of control is \$$c$/ha then control is profitable if:

$$p\ d\ l\ k > c \quad (4.3)$$

Table 4.1. *Benefits and costs of hypothetical pest control actions*

Action	Benefits	Costs	Benefits − costs	Benefits/costs
1.	$1500	$750	$750	2.0
2.	$1200	$500	$700	2.4
3.	$700	$250	$450	2.8

Norton (1976) defined the level of attack at which it becomes profitable to do control as:

$$l = c \,/\, p\, d\, k \tag{4.4}$$

The level of pest attack that corresponds to that economic threshold can be represented as shown in Fig. 4.4. The effects of year-to-year pest damage can be incorporated in the same type of analysis, as described by Norton (1976). An analysis of this type can only be done when there is an estimate of the level of attack. If that information is not available then alternatives must be used.

4.2.3 *Cost-effectiveness analysis*

An alternative analysis is cost-effectiveness, which is used when the benefits cannot be easily estimated in monetary terms, but the costs and physical returns can be estimated (McAllister, 1980). This is a common situation in vertebrate pest control, especially in conservation reserves.

An example of the analysis was reported by Hone (1990). The cost per pig of shooting feral pigs declined exponentially with increasing number of pigs shot (Fig. 4.5). A similar decline occurred with increasing kills per hour of shooting. Ridpath & Waithman (1988) reported a similar result for shooting water buffalo (*Bubalus bubalis*) in northern Australia, though they reported it in a different way. The cost per buffalo shot increased as buffalo density decreased over three years. Bayliss & Yeomans (1989) also reported a similar relationship for shooting water buffalo in northern Australia.

Cost-effectiveness analysis could also be used to compare cost-effectiveness of two or more methods of control. Taylor & Katahira (1988) compared the cost-effectiveness of aerial shooting and shooting by use of Judas goats, to eradicate feral goats in Hawaii Volcanoes National Park. Shooting aided by

Fig. 4.4. The relationships between the net revenue for a production activity with and without pest control, and pest abundance. The threshold pest abundance below which control is not profitable is *x*.

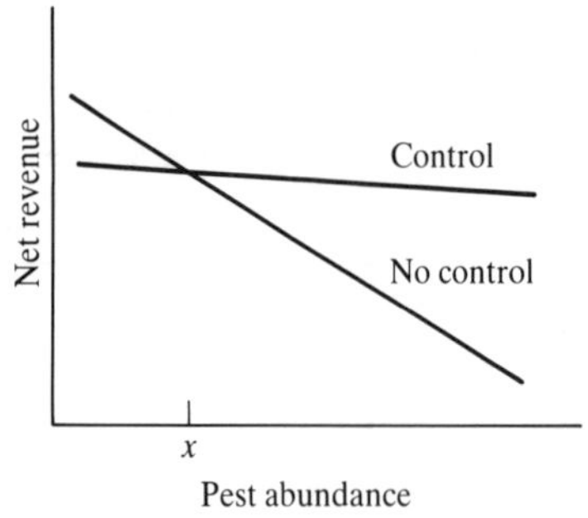

Judas goats was cheaper per goat killed, over the range of conditions encountered, because the method reduced the search time. The cost-effectiveness of shooting coyotes from a helicopter or an ultralight aircraft were compared by Knight *et al.* (1986). The ultralight had a lower cost/kill for a similar number of kills, though it tended to have more mishaps.

Theoretically, cost-effectiveness analysis could show a cross-over result when costs are assessed of two methods. For example, it may be cheaper to shoot pests when only a smaller number are killed. When a larger number are killed it may be less costly to trap. At the cross-over kill rate the cost/kill of shooting and trapping would be equal.

Choquenot (1988) estimated the cost-effectiveness of feral donkey control in northern Australia. First, the cost of instantaneous population reduction was estimated, then the cost of restraining the population at that density. The cost of killing each donkey was estimated from an inverted equation of the functional response that predicted the time, and hence cost, taken to kill each donkey (y) for a given population density (x). The growth rate of the reduced population was assumed to be logistic (see section 5.3.1) which determined the annual amount of control needed to restrain population density and hence the cost of restraint. The study predicted that the maximum cost of such control occurred at about 40% of carrying capacity ($0.4K$).

4.2.4 *Decision theory*

A fourth analysis comes from decision theory and is based on estimating the expected monetary outcome of alternative strategies. In decision theory analysis, the occurrence of events such as pest impact is not certain but the probability of their occurrence can be estimated. Hone (1980) gave an example of the application of this analysis to the control of the house mouse

Fig. 4.5. The relationship between the cost per feral pig shot and the number of feral pigs shot per square kilometre. (After Hone, 1990.)

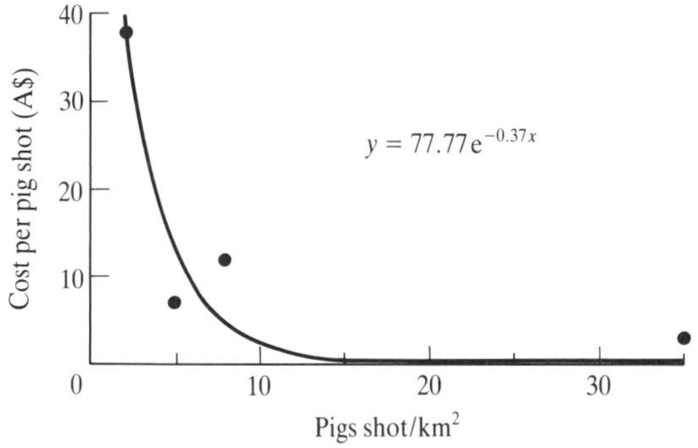

(*Mus domesticus*) (Table 4.2). In a hypothetical example a farmer has to decide to control or not control mice before the full economic impact of the mice is known. This is in fact a very common situation in many areas of vertebrate pest control. The strategy that has the highest expected profit is that which is optimal. The control strategy that maximised profit ($8285) in the hypothetical example was mouse control. The profit was estimated as $8285 = 0.715 × $8000 + 0.285 × $9000. A more detailed analysis, using logistic regression, for estimating the probabilities of plagues of house mouse in South Australia was reported by Mutze, Veitch & Miller (1990).

4.3 Spatial and temporal aspects

Tisdell (1982) examined the effects of movement by pests on the level of control. If movement occurred between private properties such as farms, then as long as farmers acted independently the level of control by each farmer would be less than the socially optimal level of control. Tisdell (1982) suggested that this occurred for two reasons: firstly each farmer does not get the full benefit of removing pests from his or her farm, and secondly, movement of pests onto a farm where control is planned will reduce the profitability of control on that farm. The co-operation of adjoining farmers in planning and doing control was suggested as being sufficient to ensure optimal control. Cherrett *et al.* (1971) noted similar points and concluded that education of farmers was necessary to address the problems.

Another spatial aspect of vertebrate pest control is that damage may be concentrated in some locations, as discussed in section 2.2. Dolbeer (1981) indicated how this may result in many locations suffering a low level of blackbird damage to corn. In these locations the cost of control exceeds the benefits.

Table 4.2. *Expected profits for a hypothetical crop in northern New South Wales for each mouse plague–control situation. The probabilities were estimated using data in Saunders & Giles (1977)*

		Strategy	
State	Probability of state	Control	No control
Plague	0.715	$8000	$6000
No plague	0.285	$9000	$10000
Expected profit		$8285	$7140

After Hone (1980).

As the spatial extent of a control area increases the cost per unit area declines. Shumake *et al.* (1979) reported such a decline for the cost of fencing areas against rodent damage. Caslick & Decker (1979) reported a similar inverse relationship between the cost of protecting trees from deer in an orchard. The cost per tree decreases as orchard area increases.

The temporal component of economic analysis can be incorporated by calculating the present value of future costs or benefits. The analysis is called discounting and is essentially the reverse of calculating interest rates (Clark, 1976; McAllister, 1980). Discounting is commonly used in cost–benefit analysis. For example, money (M_0) deposited in a bank can accumulate interest (i) at an annual compound rate. The value after n years (M_n), inflation ignored, is:

$$\text{Future value } (M_n) = M_0 \ (1+i)^n \tag{4.5}$$

The compound interest rate corresponds to the situation where the interest in each year is added to (compounded with) the initial amount of money for the calculation of interest in the next year. The emphasis here is on estimating the future value of money invested now. In discounting, the emphasis is on estimating the present value of some amount of money which will be a cost or benefit at some known time in the future. Hence, equation 4.5 is rearranged to give:

$$M_0 = M_n/(1+i)^n \tag{4.6}$$

Now the interest rate is called the discount rate. For example, the discounted value of \$10 in 15 years using an 8% discount rate is \$3.15. Conversely, \$3.15 deposited in an account with 8% compound interest per year will yield \$10 after 15 years ($\$10 = \$3.15(1+0.08)^{15}$). When the present value of costs or benefits involves more than one future time, then the present values are simply summed across each of the future times, so:

$$M_0 = \sum_{n=1}^{N} M_n/(1+i)^n \tag{4.7}$$

where the total number of years is N.

McAllister (1980) noted that the discounting procedure has an aura of precision that it does not deserve. The selection of the appropriate discount rate is a matter of much debate (McAllister, 1980), as a variation of several percentage points can be the difference between accepting or rejecting alternative plans. Clark (1976, 1981) described the results of his earlier analysis, which showed that if the renewal rate of a resource is less than the discount rate used in valuing it, then the optimal harvest strategy is to harvest to extinction. That suggests that in vertebrate pest control species with higher intrinsic rates of increase (r_m) may not be eradicated but species with lower rates may be. Economic considerations may alter such biological ideas, however. Another temporal biological consideration is replacement of one pest by another as discussed by Walker & Norton (1982). If such replacement occurs then the future costs of extra control need to be estimated.

Vertebrate pest control may produce a variable effect or response. Such variability can also be called uncertainty or risk. Managers or other people suffering pest damage may be more or less averse to risk. The effect of risk is that managers value income now more than income received at some later date. The more they value present income the more risk averse they are. The more they value future income the more risk prone they are. If they do not care they are risk neutral. In a production system, for example agriculture or forestry, the major sources of uncertainty are yield and price (Dillon, 1977). The effects of risk and how risk can be incorporated in an economic analysis were described by Dillon (1977). It was noted that the effects of risk and time were not simply additive but that an interaction can occur between them. For details of the analyses incorporating risk see Dillon (1977) and Walters (1986).

4.4 Predation control

Methods and statistical analysis for studying predation of livestock were described in section 2.4 and the statistical analysis of response to control in section 3.15.

Pearson & Caroline (1981) estimated the benefits of predator control as about 4.5 times the costs, for 21 counties in central Texas. The predators were coyotes and bobcats and losses were of sheep and goats. The control methods were not clearly stated but presumably included trapping. The economic evaluation requires close examination. The benefit/cost ratio was estimated from the losses in the absence of control ($1 169 000). The livestock losses when predator control occurred were estimated as $29 300. The difference ($1 139 700) is the reduction in losses because of control, and is the benefit from control. The costs of control were the salaries and expenses which were estimated to be $231 689, plus predation losses of $29 300, giving a total of $260 000. Hence, an estimate of the benefit to cost ratio is $1 139 700/$260 000 = 4.4:1.

The costs and benefits of using guard dogs to control predation of livestock were estimated by Green, Woodruff & Tueller (1984). Sheep producers were surveyed by questionnaire. Costs included the maintenance and training of the dogs and the benefits were the reduction or elimination of losses. Green *et al.* (1984) estimated the practicality of guard dogs by subtracting the costs from the benefits for each producer. For over 27 producers that had higher benefits than costs this averaged \$2554. Ten producers regarded the dogs as a liability as the costs exceeded the benefits by \$666 per farm. The benefit/cost ratio was not estimated but using the data of Green *et al.* (1984) the ratio was 5.2:1 (=\$3836/\$739) over all producers.

Smith, Neff & Woolsey (1986) studied the benefits and costs of controlling coyotes to increase the hunting harvest of pronghorn (*Antilocapra americana*) in Arizona. Trends in abundance of coyotes and pronghorn were estimated by aerial survey. The predicted effects of eight strategies for coyote control were studied by computer simulation. The costs of control were estimated from field data on trapping and shooting, the latter from a helicopter. Hunting permits were used to generate an estimate of economic benefit through gross expenditure on hunting. Net benefits were estimated as benefits minus costs summed over the ten year period of the computer simulation.

All control strategies were compared with that involving coyote control in the first year only (the default strategy). The strategy with the highest net benefit (1.92), relative to the net benefit of the default strategy, was coyote control every second year (Table 4.3). That strategy had the highest benefits (\$519 981) but not the lowest costs (\$34 400) which occurred with coyote control every five years. Although this was not calculated in the paper, the default strategy had the highest benefit/cost ratio (14.2:1) (Table 4.3). The strategy of control every second year had a ratio of 6.1:1.

Predation of lambs by feral pigs may not occur in each lambing (Pavlov *et al.*, 1981). A payoff matrix may be a suitable analysis for estimating the consequences of pig control when control occurs at the start of lambing prior to the evidence that predation occurred. Using data from Pavlov *et al.* (1981) the expected outcomes of doing or not doing control can be described and are listed in Table 4.4.

Table 4.3. *Estimated benefits and costs of selected coyote control strategies*

Control strategy	Benefits	Costs	Net benefits	Net benefits ratio	Benefits/ costs ratio
Default (in year 1 of 10)	\$243 507	\$17 200	\$226 307	1.00	14.2:1
First of 2 years	\$519 981	\$86 000	\$433 981	1.92	6.1:1
First of 5 years	\$320 356	\$34 400	\$285 956	1.26	9.3:1
First 3 of 5 years	\$517 203	\$103 200	\$414 003	1.83	5.0:1

After Smith *et al.* (1986).

The expected outcome of doing pig control is 68 lambs per 100 ewes. This assumes that pig control will prevent 90% of predation, and is estimated as $68 = 0.5 \times 65 + 0.5 \times 70$ lambs (rounded to the nearest whole number). The expected outcome of no control is estimated as 44 ($= 0.5 \times 18 + 0.5 \times 70$) lambs. Given the uncertainty about whether predation will or will not occur, the strategy that maximises the number of lambs is to do control. The sensitivity of the analysis can be examined by varying the estimated proportional reduction in predation following pig control. Obviously, if control eliminates predation then the expected outcome is 70 lambs ($= 0.5 \times 70 + 0.5 \times 70$). Iterative analysis shows that as long as control saves at least two lambs then the expected outcome will favour control. For example, if control occurs and the number of lambs produced is 20 then the expected outcome is 45 lambs ($= 0.5 \times 20 + 0.5 \times 70$). The economic analysis of this situation would include the costs of control relative to the value of the lambs produced. The estimated costs should allow for more than the cost of pig control. If lambs are saved, then there will be extra costs incurred in getting them to market, such as drenching, maybe shearing and marking.

Mathematical modelling of predation of livestock is briefly discussed in sections 5.5 and 6.7.

4.5 Control of infectious diseases

Statistical analysis of damage caused by infectious diseases was discussed in section 2.5 and analysis of disease control in section 3.16.

As a sweeping generalisation, there have been remarkably few economic analyses of the control of infectious diseases involving vertebrate pests. The reasons for this are unknown.

Anderson *et al.* (1981) considered that control of rabies in European foxes

Table 4.4. *Payoff matrix of the number of lambs produced per 100 ewes, if control or predation by feral pigs do or do not occur*

		Strategy	
State	Probability of state	Control	No control
Predation	0.5	65	18
No predation	0.5	70	70
Expected outcome		68	44

Data are based on those reported by Pavlov *et al.* (1981).

may be achieved in the most cost-effective manner by a combination of vaccination and culling, though in the absence of cost data they did not do an economic analysis. Coyne, Smith & McAllister (1989) estimated the costs of culling, vaccination, or a combination of both, to eliminate rabies in raccoons. The analysis used was linear programming, which is a technique that solves a linear equation that shows the costs of both culling and vaccination. Coyne *et al.* (1989) assumed that the costs were estimated by the equation:

$$Z = cA + vB \tag{4.8}$$

where Z is the cost of the disease elimination programme, A is the cost of culling, B is the cost of vaccination and c and v are coefficients. The aim of the analysis was to minimise the value of Z. The linear equation is equivalent to assuming a constant marginal cost of culling and vaccination. This is unlikely, as noted by Anderson *et al.* (1981). It was assumed that the coefficients c and v were limited between zero and one, so the value of Z was actually an index of the total cost. The result of the analysis was that the lowest cost strategy was always either culling or vaccination used alone.

The accumulated costs, discounted and non-discounted, of culling, vaccination and fertility control, singularly or in combination, for tuberculosis control were reported by Barlow (1991b). The tuberculosis occurred in brushtail possums in New Zealand. Costs were accumulated over eight years, the assumed time needed to eliminate tuberculosis. The benefits of such tuberculosis control were not estimated.

Mathematical modelling of the dynamics of infectious diseases is described in section 5.6 and control of infectious diseases in section 6.8.

4.6 Rodent damage control

Statistical analysis of rodent damage was discussed in section 2.6 and analysis of damage control in section 3.17.

The literature on the economics of rodent control is limited and appears to be several orders of magnitude less than the literature on techniques of rodent control. Richards (1988) provides several interesting examples of cost–benefit analysis of rodent control.

Saunders & Robards (1983) estimated the economic return to a farmer of control of house mouse in a sunflower crop in southern New South Wales. The broad steps in the study were as follows. The mouse population density was estimated by mark–recapture as 2716/ha, seven weeks prior to harvest. Using the method of Dolbeer (1975), the damage to the sunflower crop was estimated to be 12.4% at the time of population estimation. If the mouse population remained stable till harvest and if each mouse daily consumed 3.4 g of sunflowers then the estimated additional loss was 22.6% of expected average yield. Hence, in the absence of control the yield reduction was predicted to be 35% (=12.4+22.6).

Poisoning, shortly after population estimation, lowered mouse abundance by up to 90%. A subset of the study compared the effects of three pesticides on mouse abundance. Assuming that the lowered mouse population ate sunflowers at the same daily rate the yield reduction caused by the lowered mouse population would be 2.2%, instead of the 22.6% loss predicted in the absence of control. This is the maximum effect of the pesticides as it assumed a 90% reduction in mouse abundance. Hence, the benefit of control was the saving of the reduction in yield of 20.4% (=22.6−2.2).

Saunders & Robards (1983) estimated that the net benefit (benefits−costs) of mouse control was A$123/ha. The benefit/cost ratio, though not calculated by the authors, was 16.4:1. The calculations were recognised as simplistic but provide a useful guide. A crude extrapolation suggests that the savings would be higher if the effect of pesticides was higher or if control occurred earlier. The former could be complicated by the need to estimate the marginal benefits and costs of the increased pesticides. The latter could be complicated by breeding and immigration resulting in the need for retreatment, as noted by the authors.

An evaluation of poisoning for control of damage by black-tailed prairie dogs (*Cynomys ludovicianus*) in South Dakota estimated that control was not economically sound (Collins *et al.*, 1984). The study was unusual for the detailed methods used and that control had occurred for over 80 years apparently without such an economic study. The broad steps in the study are described. Estimates were made of the increased forage available to cattle after prairie dog control, though statistical analysis reported no significant difference in total biomass. The extra forage was converted to estimates of the numbers of extra animal grazing units that the land could then support. The extra animal units were used to assess economic benefits.

The costs of control were estimated for two control scenarios: firstly, annual maintenance to prevent repopulation of the control area, and secondly, complete retreatment once the prairie dogs had repopulated the entire area. The area needing to be treated in annual maintenance control was estimated at 5, 10, 20 or 30% of the area initially treated. Complete retreatment was predicted to be needed every 3, 6, 9 or 15 years. The first year of control was used as the base year for discounting future benefits and costs. All benefits and costs were discounted using a 4% discount rate and estimates were free of inflation. The annual net benefits were estimated separately for ranchers and for the U.S. Forest Service, as the costs and benefits were slightly different for each.

Only annual maintenance control on less than 10% of the area could recover the costs of the initial control program. If 5% of the area had to be annually treated then it took 40 and 22 years for the Forest Service or ranchers, respectively, to recover initial control costs. If complete retreatment of the area was needed then the initial control costs could never be recovered. Subsequent research (Knowles, 1986) reported that two populations of prairie dogs had recovered, or nearly so, to pretreatment levels after five years, so only the three year retreatment analysis used by Collins *et al.* (1984) appears relevant.

The difference between the benefits and costs of control of house mouse in soybean crops in western New South Wales was estimated by Singleton *et al.* (1991a). The benefits were estimated as the difference in damage between a crop where mice were poisoned and a crop where they were not poisoned, multiplied by the monetary value of such damage. Costs were for materials and labour. The difference between benefits and costs was positive in one crop, but changed over time from positive to negative in the second crop. It was considered that house mouse control was economically marginal, though a full analysis was not feasible. It was concluded that the economic analysis was limited by a strong effect of crop density on the level of damage, from which the effect of poisoning could not be separated.

Mathematical modelling of rodent damage and dynamics is briefly addressed in sections 5.7 and 6.9.

4.7 Control of bird strikes on aircraft

Bird strikes on aircraft undoubtably can cause damage to the plane. The monetary value of such costs can be estimated and may be measured in thousands or millions of dollars or pounds (Murton & Westwood, 1976). The net monetary value of any control measures could also be estimated, with difficulty, by valuing the benefits and costs. There is scope for further estimation of such benefits and costs.

4.8 Control of bird damage to crops

The statistical analysis of physical and production effects on crops of blackbirds and other bird pests was described in section 2.8. Statistical analysis of control was discussed in section 3.19.

The damage by wood-pigeons to vegetable crops and costs of control in parts of England were compared by Murton & Jones (1973). They expected a negative correlation between the amount spent on crop protection and the amount of damage. In contrast, on farms in one area there was a significant positive rank correlation but in a second area there was no significant correlation. Hence, it appears that at the first site the control may have been in response to damage but the control had no effect on the amount of damage.

Dolbeer (1981) described a method of using estimates of the benefits and costs of blackbird control. The simple assumption was that the benefits must exceed the costs of control. The benefits were estimated as the value of the reduction in damage. Hence, the analysis was a simplified version of cost–benefit analysis, with no temporal effects and hence no discounting. The method involved estimating the relationship, as a regression, between the benefits and costs. They could be graphed as shown in Fig. 4.6. The regression line for method 1 in Fig. 4.6 corresponded to the break-even point where the benefits equalled the costs. A level of control that estimated benefits above the line resulted in the benefits exceeding the costs. In Fig. 4.6, for any level of benefits method 1 has lower costs than method 2.

The analysis was applied to data on blackbird damage to corn crops in Ohio (Dolbeer, 1981). Of 21 fields studied, two had levels of damage that could have been controlled economically. The control method was apparently ineffective. The results suggest blanket control would be uneconomic and so identification of fields with most damage would be of great assistance. This is simply recognising the spatial variation in damage discussed in Chapter 2. Which corn fields suffered the highest damage was apparently inversely related to the distance from a blackbird roost.

The analysis was recognised by Dolbeer (1981) as simplistic. Points not incorporated in the analysis were possible beneficial effects of birds and externalities of control. The analysis was, however, an advance on guesses or on the uniform application of control to all corn fields. It could be expected that the relationship shown in Fig. 4.6 would be curvilinear as marginal benefits

Fig. 4.6. The expected relationships between the benefits and costs of blackbird control for two methods of control. (After Dolbeer, 1981.)

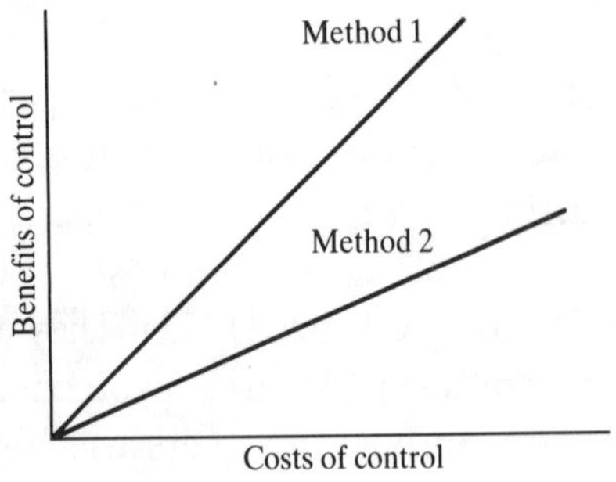

decrease with increasing costs of control as shown in Fig. 4.3. At low levels of control the linear and curvilinear relationships are probably nearly identical, so the difference may be of theoretical interest only. Also, the analysis estimated whether control was profitable but did not estimate the most profitable control level.

The economic pest status of quelea in Africa was reviewed by Elliott (1989), who considered the supporting evidence for its pest status was scanty. Data on the levels of damage to crops at the local, national and continental levels were collated, though many damage estimates were in physical production units (tonnes, %, ha) rather than monetary units ($). It was concluded that quelea damage probably represented less than 1% of total cereal production, though locally there may be high economic costs. There were no data on the benefits and costs of control of such damage, though it was noted that in parts of Africa the lost production from quelea damage is not lost income but lost food. Bruggers (1989) reported some estimates of benefits and costs of quelea control in east Africa. Control was by use of chemical repellents. Costs were estimated as the sum of chemicals and application, and the benefits as the yield difference between sprayed and unsprayed crops. The latter were as simultaneous or sequential experimental controls.

Mathematical modelling of bird damage to crops is discussed in sections 5.9 and 6.11.

4.9 Rabbit damage control

Statistical analysis of rabbit damage was discussed in section 2.9 and analysis of rabbit damage control was discussed in section 3.20.

Fennessy (1966) considered that rabbits were the most important herbivore competing with livestock in Australia; however, he noted that there were little quantitative data to show the effects. That statement is still true.

One example of the economic impact of rabbits and their control cited by Fennessy (1966), and later by Cooke & Hunt (1987), was a 26% increase in sheep numbers and a 2% increase in wool cut per sheep after almost complete eradication of rabbits from a farm in South Australia. The production statistics were calculated from a seven-year period prior to rabbit control and the seven-year period after control. Weather conditions were reportedly less favourable after rabbit control so the difference was not simply associated with rainfall and pasture conditions.

Data reported by Cooke (1981) on rabbit numbers in a part of South Australia and the costs of control can be analysed to investigate the relationship between the two variables. Increasing numbers of rabbits after control were associated with decreasing costs (Fig. 4.7). Each point corresponds to a different control method as described in Cooke (1981). A linear regression of costs and abundance was significant with a correlation coefficient of -0.94 (df$=6$, $P<0.01$). A negative exponential regression was also significant with a slightly higher correlation coefficient of -0.96 (df$=5$, $P<0.01$). The latter regression had one less degree of freedom, as one data point, for no control, could not be used as the costs were zero and one cannot enter the natural logarithm of zero into a calculator or computer. The negative exponential regression is the same trend as shown in Fig. 4.2.

Although Cooke (1981) used an elegant experimental design and analysis of variance to estimate the effects of control on rabbit abundance, it was noted that farmers were more interested in reductions in crop damage. In contrast to the elegance of the statistical analysis, damage estimates were obtained in the study by subjective scoring on a four point scale from zero to severe.

Foran *et al.* (1985) concluded that rabbit control in parts of central Australia would not produce an economic return, in the short-term. Off-farm investment would probably yield a higher rate of return on investment; that is, produce greater monetary benefits.

The economic value of changes in crop and grazing production after the occurrence of myxomatosis in Australia and Britain was described by Sumption & Flowerdew (1985). Each change involved a simple sequential comparison of yields before and after myxomatosis. The associated losses in sporting and fur industries were noted but not expressed in monetary terms.

Fig. 4.7. The regressions between the cost to reduce rabbit numbers to a certain level and the number of rabbits after such control. The linear regression ($y=61.00-0.64x$) is the solid line and the negative exponential regression is the dashed line. Calculated from Cooke (1981) using the mean number of active entrances in April on the x axis. Each data point is a different combination of control methods, including no control.

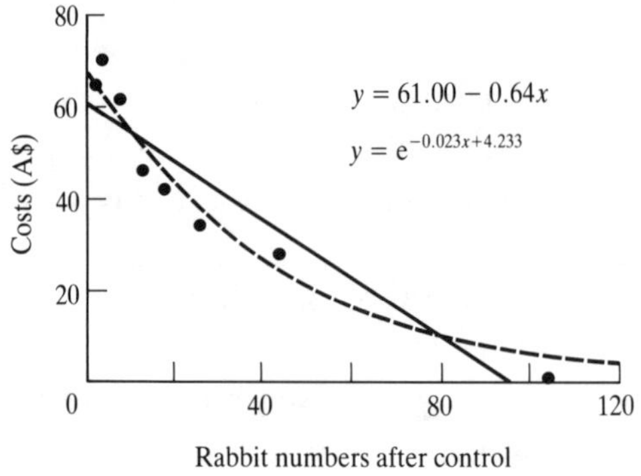

Sheail (1991) however was critical of the limited data on the economics of rabbit control in Britain.

Mathematical modelling of rabbit damage control is discussed in sections 5.10 and 6.12.

4.10 Conclusion

The economic analyses that could be applied far exceed their application to vertebrate pest control. It is my opinion that this situation has developed because of lack of effort rather than lack of relevance of analyses or the analyses being difficult to apply. The case studies described are not always supportive of pest control. This suggests a greater need for economic analyses to determine where control is profitable. The amount of data needed for an economic analysis can be large (Table 4.5).

The literature on different topics in this chapter is very unbalanced. For example, there is some data on the economics of predation control, and little on control of infectious diseases and bird strikes. These conclusions will of course be biased if I have missed the relevant literature, but if I've missed it so will lots of others.

There is scope for further economic analysis of the use of two or more control methods. Some research has occurred on the relative benefits and costs of alternative control methods. There appears to be little or no research on the economics of partly substituting one control method for another and hence estimating the marginal rates of substitution (Dillon, 1977).

Table 4.5. *Data needed for economic analysis of the benefits and costs of vertebrate pest control*

1. *Benefits*	– increase in returns and costs foregone
	– total benefits
	– marginal benefits
	– probability that benefits will occur
	– present value of discounted benefits
	– spillover benefits
2. Costs	– opportunity costs and increases in things not desired
	– total costs
	– marginal costs
	– present value of discounted costs
	– spillover costs such as social and non-target effects

The analysis of the need for control and of the level of pest damage can be assisted by mathematical models of the dynamics of pest populations and the damage the pests cause. That is the subject of the next chapter.

5

Modelling of populations and damage

A model represents an event or process in a different and usually simplified form. Modelling, like statistics, can evoke great passions from the uninitiated or strongly prejudiced. To others it's like trying to understand cricket when you have previously played baseball. For readers that find mathematics very trying, a very readable discourse on mathematics and science by Kline (1953) is recommended to help break the ice, and for a similar discourse on modelling see Walters (1986).

Krebs (1988) suggested that population ecologists did not know enough to construct useful models of rodent population dynamics, though Caughley & Krebs (1983) modelled plant–herbivore dynamics of small and large mammals. A good point raised by Krebs (1988) was the need to bridge the apparent gap between modellers and field ecologists by describing the types of data to be measured in the field. Too often, field ecologists collect data not relevant to a model and modellers use parameters in a model that can not be directly estimated in the field. Hairston (1989) correctly cautioned against the blind acceptance of mathematical models because of their inclination to make unrealistic assumptions. Krebs (1988) doubted if models produce new principles in ecology. That is probably true of many models, but the models of disease dynamics have generated the concept of a threshold host abundance. This is described in section 5.6.

This chapter is about modelling the processes of changes in pest abundance and damage of vertebrate pests. Some of the ideas have been published widely and others are new. A common theme through the chapter is the analogy between events and processes in vertebrate pest control and in theoretical ecology. Emphasis is on graphical and mathematical models. The chapter describes, in sequence, models of population growth of large then very small populations, models of pest and disease movement, then of damage and infectious diseases. In this and Chapter 6 many of the steps are shown in deriving a model or drawing out the practical implications of a model. I encourage readers to follow the steps to get a better knowledge and understanding of the origins of the conclusions.

5.1 Uses of models

Models are used to: (i) describe the factors that influence an event or process and how those factors interact, (ii) identify strategies for control of damage or populations, (iii) predict the occurrence of damage and response of damage or populations to control. Conway (1977) and Conway & Comins (1979) considered that the main use of mathematical models in pest control is to provide guidelines for evaluating alternative control strategies. If the models' parameters can be estimated then the models may also be of predictive value. Strategic models are intended primarily to give broad guidelines on how a particular pest problem should be studied or tackled. Tactical models, by contrast, are more specific for a particular pest problem at a certain time and place (Conway, 1984). Both types of model are described here.

Models of pest control can be tested in two ways: firstly, by comparing predictions with observed data, and secondly, by testing one or more assumptions in the models. As a sweeping generalisation, there is a great need for both tests. Guidelines for testing of models are described in the conclusion of the chapter.

5.2 Types of model

Models can be sets of thoughts (conceptual models), graphs or mathematical equations. Thought or conceptual models are of the verbal or written kind where ideas are expressed but not graphically or mathematically. Graphical models represent ideas in graphs or figures, but without describing the ideas or trends by equations. Mathematical models use equations and can be deterministic or stochastic. The former predict one result for any one input and the latter predict a set of results for any one input, with some results more likely than others.

In this chapter, and the next, simplified deterministic models rather than more complex models are described. This is done to help those with limited experience of modelling, yet is recognised that the approach can be criticised as overly simplistic. However, as Maynard Smith (1974) and Walters (1986) describe, simple models are easier to understand and evaluate, without great loss in accuracy. An alternative justification is to apply Ockham's razor and adopt the simpler rather than the more complex model, assuming each has equal power of prediction.

5.3 Modelling pest population dynamics

There are many models of wildlife population dynamics. This section will not attempt to describe or review all of those, rather it will describe those of most relevance to vertebrate pest control. The models are particularly relevant for large populations. There is no hard and fast rule about how big large is, but as a working guide I have chosen a minimum of about 100 individuals in a population. Most models assume that the populations are closed – no immigration or emigration, and many examine trends over years so do not describe seasonal trends in abundance.

5.3.1 *Predicting pest abundance*

These models are concerned with predicting population size at some time in the future. Models of population growth are examined firstly for single species, hence one trophic level, then for two or more trophic levels. The simplest models are examined first then more complex models. The models relate abundance to prior abundance, time and rate of increase. Demographic parameters such as age structure and sex ratio are not included in such models.

The simplest assumption for growth of a pest population is exponential growth (Fig. 5.1a). Expressed as a difference equation this is:

$$N_t = N_0 \, e^{rt} \tag{5.1}$$

$$= N_0 \, \lambda^t \tag{5.2}$$

Fig. 5.1. Growth of pest populations. (*a*) Exponential growth as described by equations 5.1 to 5.3 when the exponential rate of increase (*r*) is positive. (*b*) Logistic growth as described by equations 5.4 to 5.7 when the exponential rate of increase is positive. Carrying capacity is *K* when growth is logistic.

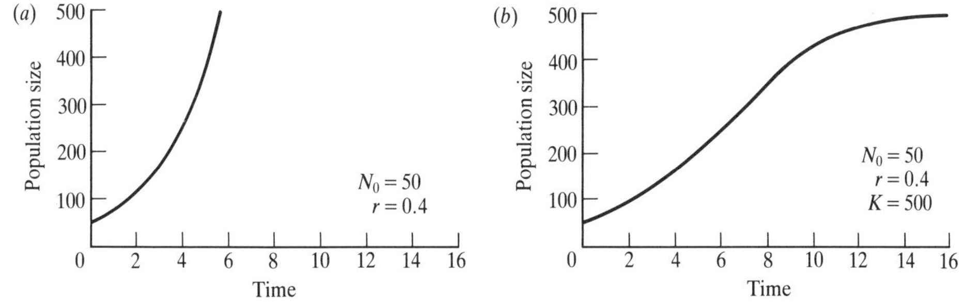

where N_t is the population size at time t and N_0 is the population size at time 0. The exponential rate of increase is r and the time is t. The finite rate of increase (λ) is equal to e^r. The maximum value of r is r_m which is the intrinsic rate of increase. Statistical analyses for estimating λ were examined by Eberhardt & Simmons (1992) who reported little difference in bias but confidence limits were underestimated.

Expressed as a differential equation, exponential growth is:

$$dN/dt = r\ N \tag{5.3}$$

The population has no upper size. This is obviously simplistic as populations do not increase indefinitely. Hone (1977) reported one example of exponential growth of a population, but the dynamics of the population were anything but simple and the model had, at best, imaginary links to wildlife ecology. McCallum & Singleton (1989) assumed exponential growth of a model population of house mouse.

The next step up the ladder of population growth realism is logistic growth (Fig. 5.1b). Expressed as a difference equation (Eberhardt, 1988) this is:

$$N_t = N_0 + r_m N_0 (1 - (N_0/K)) \tag{5.4}$$

where K is the carrying capacity. Logistic growth expressed as a differential equation is:

$$dN/dt = r_m\ N\ (1 - (N/K)) \tag{5.5}$$

$$= r_m\ N\ ((K - N)/K) \tag{5.6}$$

The terms in the logistic equation can be estimated using methods described by Caughley (1980). Integration of the differential equation gives an equation for logistic growth, one form of which (Krebs, 1985) is:

$$N = K/(1 + e^{a - rt}) \tag{5.7}$$

where a is a constant.

The logistic equation can be rearranged to show other aspects of dynamics. The per capita rate of increase ($(dN/dt)/N$) is shown in Fig. 5.2 and expressed mathematically as:

$$dN/dt/N = r_m\ (1 - (N/K)) \tag{5.8}$$

$$= r_m - r_m\ N/K \tag{5.9}$$

The equation states that when $N = K$, $dN/dt/N = 0$, so the population is stable, and hence not increasing or decreasing. As N approaches zero, N/K approaches zero, so $dN/dt/N$ approaches r_m. Equation 5.9 is a linear regression with the independent variable ($x = N/K$), a slope of $-r_m$, and an intercept of r_m as shown in Fig. 5.2. The relationship is an assumption of logistic growth (Krebs, 1985) that can be tested. The other assumptions are that the population has a stable age distribution, density (or abundance) is measured in appropriate units, and density-dependent effects act instantaneously without any time lags.

The logistic model of population growth was used by Caughley (1980) for general investigations of vertebrate pest control, by Anderson *et al.* (1981), Murray, Stanley & Brown (1986) and Murray & Seward (1992) to model dynamics of foxes in Europe, and by Pech & Hone (1988) and Pech & McIlroy (1990) to model dynamics of a feral pig population in Australia. Clout & Barlow (1982) also used a logistic model of population growth to estimate the maximum sustained yield of possums (*Trichosurus vulpecula*) in New Zealand as part of an investigation of control compared with harvesting of possums. Wood & Liau (1984) fitted a logistic growth equation to the increase in numbers of *Rattus tiomanicus* in an oil palm plantation in Malaysia. In the latter study, in one time period, October 1972 to April 1974, the correlation coefficient for logistic growth was $r = 0.84$, but for a linear increase was higher ($r = 0.94$). Both coefficients would have been statistically significant ($P < 0.05$), though that was not cited in the paper. In two other time periods the fit to logistic growth was slightly higher than that for linear growth. Choquenot (1988) used the logistic equation to model growth of a feral donkey population in northern Australia, as discussed in section 4.2.3.

Fig. 5.2. The relationship between per capita rate of increase of a population showing logistic growth and population size as described by equations 5.8 and 5.9. Maximum (intrinsic) rate of increase is r_m and carrying capacity is K.

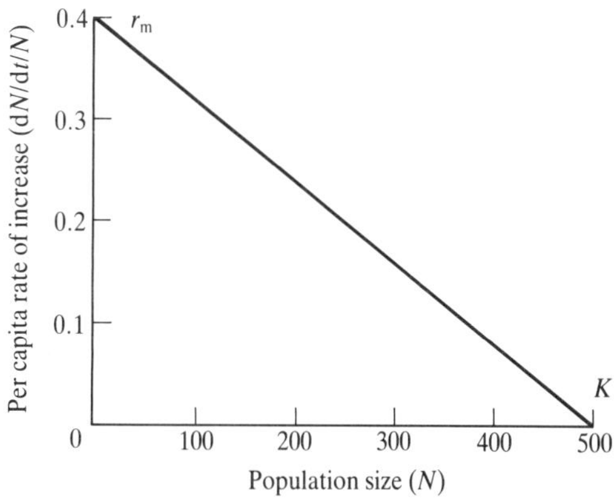

Logistic growth predicts that the net recruitment between time periods (dN/dt) is zero when the population size is zero and when it is at carrying capacity (K). Net recruitment is highest when population size is at half carrying capacity ($K/2$) (Fig. 5.3), as found by solving the quadratic version of the logistic equation which can be obtained by multiplying through by $r_m N$ in equation 5.5.

A variation on logistic growth is the generalised logistic (Fig. 5.3). This equation reflects the observation in some large mammals (Fowler, 1988) that density-dependent growth is most severe near carrying capacity (K). The difference version of the equation is:

$$N_t = N_0 + r_m N_0 \, (1 - (N_0/K)^z) \tag{5.10}$$

where z is a measure of curvilinear density-dependent effects on the ratio of N_0 to K. When $z = 1$, the equation reduces to the logistic model (equation 5.4). For $z > 1$ the density-dependent effects are most severe at high density.

Barlow & Clout (1983) examined the generalised logistic and its application to management of possums (*Trichosurus vulpecula*) in New Zealand. They had no empirical estimates of z, but examined values of 1, 2, and 4, and after using a form of sensitivity analysis concluded that the simple logistic equation was a useful starting point for estimating harvest rates but that the harvest could be modified if the empirical evidence suggested that density-dependent mortality was better described by a generalised logistic. Any errors in estimated harvest

Fig. 5.3. The relationships between net recruitment and population sizes for a population increasing according to logistic growth (solid line), and increasing according to the generalised logistic model as described by equation 5.11. The solid line shows growth for $z = 1$ (logistic growth), and the other lines are for $z = 3$ (-----) and $z = 7$ (·········).

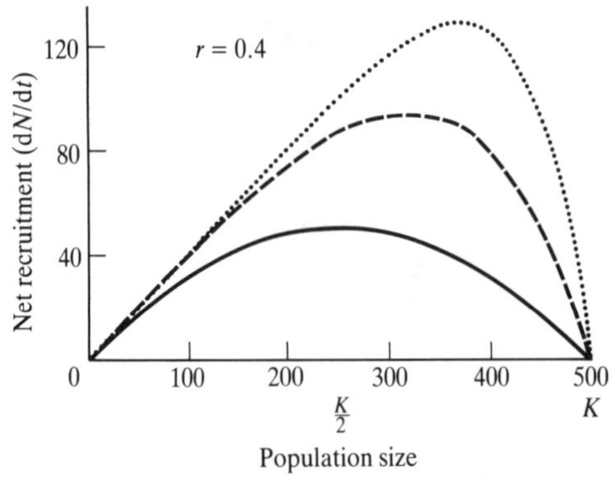

rates would be higher when z is higher. The implication for pest control is that a population has a higher net recruitment, than in simple logistic growth, so a greater control effort is needed to reduce abundance (Barlow & Clout, 1983).

Anderson & Trewhella (1985) estimated that $z = 7$ for badgers (*Meles meles*) based on population recovery data from England and Czechoslovakia. The z value was selected after iterative fitting of predicted curves to observed data. No statistical measures of goodness of fit were cited. Eberhardt (1987) chose a value of $z = 11$ to obtain agreement between observed and expected trends of elk (*Cervus elaphus*) in Yellowstone, U.S.A. The criterion for selecting the z value was to minimise the Chi-square statistic estimated by calculating the difference between observed and predicted data. Barlow (1991a,b, 1993) assumed that $z = 3$ for brushtail possums in New Zealand.

The differential equation version of the generalised logistic equation is:

$$dN/dt = r_m N_0 \, (1 - (N_0/K)^z) \tag{5.11}$$

In logistic growth, net recruitment increases with increasing population size up to a point (p), the inflection point, then net recruitment starts to decrease, returning to zero when population size equals carrying capacity (Fig. 5.3). The inflection point in logistic growth occurs at $p = K/2$. The inflection point in generalised logistic growth (Eberhardt, 1987) occurs at:

$$p = (1 + z)^{-1/z} \tag{5.12}$$

Hence, when $z = 1$, $p = 0.5K = K/2$ and when $z = 11$, $p = 0.8K$ (Fig. 5.4).

Fig. 5.4. The relationship between the inflection point (p), as a proportion of carrying capacity (K), and the exponent (z) in generalised logistic growth, as described by equation 5.12.

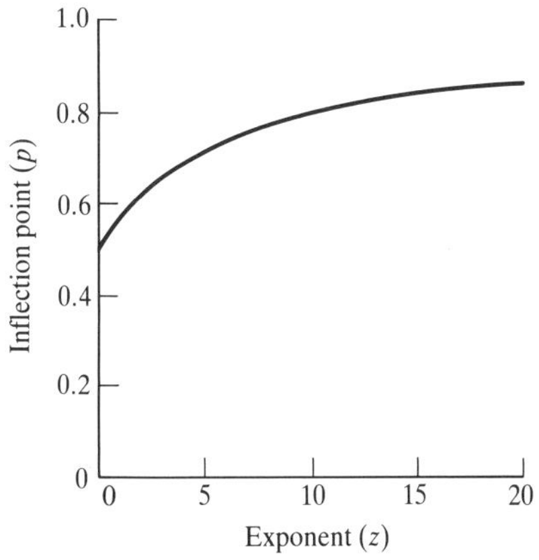

May (1981a, 1986) described a variety of single-species models of population growth. The predicted trends in abundance ranged from stable equilibria, to stable cycles to apparently chaotic patterns, depending on several population parameters. The models were often for populations with non-overlapping generations, which means they are not directly relevant for vertebrate pests, but they are of general interest. The models are simple but ignore the other components of any ecosystem in which animals occur. For example, not all the models directly address population interactions such as interspecific competition or predator–prey relationships (May, 1986). Barlow & Clout (1983) also described a variety of single-species models.

Population size or rate of increase is expected to be related to the environment of the pest, its food supply and its predators. Such relationships can be incorporated in more realistic models of population growth. Several studies have reported significant relationships between the annual intrinsic rate of increase of a population and average body weight in mammals, including Caughley & Krebs (1983), Hennemann (1983), Anderson & Trewhella (1985) and Robinson & Redford (1986). Anderson & Trewhella (1985) used the relationship to estimate the intrinsic rate of increase of a population of badgers in England, and Coyne *et al.* (1989) used the equation of Hennemann (1983) to estimate the intrinsic rate of increase of raccoons in the eastern U.S.A. The regression may be of limited value for species that have been recently introduced to an area as the absence of natural predators and pathogens may result in a higher intrinsic rate of increase than in the natural range of the species. An alternative method of modelling rate of increase is to estimate it from the age structure and fecundity schedules of a population using Lotka's equation (Caughley, 1980). Some such applications have been inappropriate as they implicitly assumed the rate of increase was zero (Caughley, 1980).

One approach to modelling rate of increase is to assume that the change in, for example, herbivore population size is related to rainfall, usually prior rainfall, as described by Myers & Parker (1975) and Foran *et al.* (1985) for rabbits, Woodall (1983) for feral pigs, Bayliss (1985, 1987) for kangaroos, Briggs & Holmes (1988) for ducks and Freeland & Boulton (1990) and Skeat (1990) for feral water buffalo. The analyses differed between studies but the approach was similar. The basic idea is shown in Fig. 5.5. Rainfall is used as an index of food supply, or maybe breeding sites, as it is much easier to measure. The rate of increase of the population is negative when rainfall is low, and has an upper limit, the genetically set intrinsic rate of increase (r_m), corresponding to limits on rates of breeding. The rate of increase may be more explicitly linked to food supply in detailed studies as reported by Bayliss (1987) for kangaroos. The relationship in Fig. 5.5 is the numerical response originally described for predator–prey relationships. In general terms the above approach may be more relevant to herbivore populations in arid and semi-arid environments, than in higher rainfall environments.

One theoretical difficulty with the numerical response relationship shown in Fig. 5.5 is that a pest population during a period of low to very low rainfall may decline much faster than it could subsequently increase. Hence, the assumption of reversibility, implicit in the numerical response, may need to be changed. A population may follow one trajectory during rapid decline but a different trajectory during a subsequent increase, that is, like a hysteresis loop.

The abundance of a population may be determined by density-dependent mortality caused by food shortage. This is the food hypothesis of population regulation. Sinclair, Dublin & Borner (1985) tested the hypothesis and provided evidence to support it for Serengeti wildebeest (*Connochaetes taurinus*). Sinclair *et al.* (1985) noted that such density-dependent mortality could also be caused by predation or disease but considered that in their study such was unlikely. Wildebeest are not vertebrate pests but the conclusions are relevant to large mammals that are. Freeland & Choquenot (1990) and Choquenot (1991) provided further support for the food hypothesis from their studies of feral donkeys (*Equus asinus*) in northern Australia.

The relationship between rainfall and pest abundance may be more complex than that shown in Fig. 5.5. For example, Singleton (1989) reported that a plague of house mouse in southern Australia in 1984 appeared to follow above-average autumn rainfall about 15 months earlier, followed by high winter and summer rainfall. The autumn rain had followed a drought. Hence, the plague occurrence was consistent with the drought–plague hypothesis proposed by Saunders & Giles (1977). Neither study presented the relationships as mathematical models but Singleton (1989) presented his results as a

Fig. 5.5. A numerical response relationship between rate of increase of a herbivore population and rainfall.

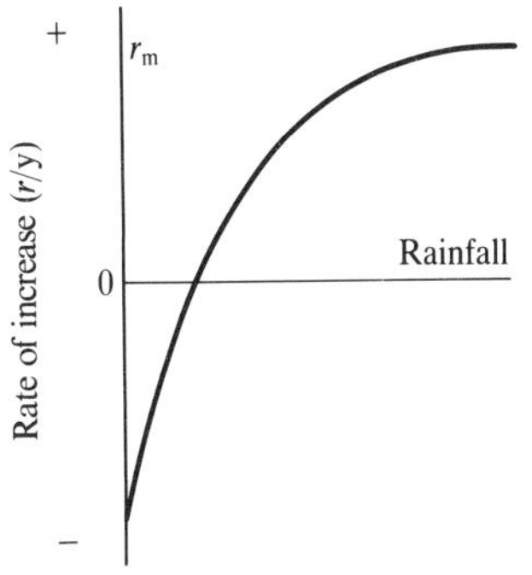

graphical model as did Redhead (1988) for a different but related model. Mutze *et al.* (1990) reported an empirical model, using logistic regression, of plagues in South Australia. The key variables used in the predictive model were measures of rainfall and a measure of the difference in yield of wheat crops in different years. Hence, the model used measures of both climate and food supply.

The effect of environment on population growth may not always act through rainfall. Gosling, Baker & Skinner (1983) described a simulation model of a coypu population in south-eastern England. The effects of temperature were included in the model. It was assumed that the effects of rainfall were insignificant. The model estimated the number of females in the population each month. Males were ignored. The model also estimated the number of females born each month, the number of young which matured to be adult females, and the number of coypu killed by trapping each month. Reproductive output was linked to a measure of winter severity formulated by adding the number of successive days with minimum temperatures below or equal to 0 °C and maximum temperatures below or equal to 5 °C. The computer simulation was run for continuous sequences of cold winters and separately for mild winters. The predicted number of females was generally higher with mild winters. When the sequence of winter temperatures was randomised between years, the predicted results differed with starting population size and trapping effort.

Multi-trophic models of population involve predators, prey and maybe prey food. Vertebrate pests may be either the predators or the prey. Coyotes are clearly predators and livestock the prey. Rabbits are a predator of grass (the prey) and may themselves be prey for foxes. The general structure of multi-trophic models can be based on the Lotka–Volterra models (May, 1981b). Models of a simple predator–prey system have two equations, linked by predator offtake of prey:

Change in prey population = growth of prey in absence of predator − number of prey eaten by predators (5.13)

Change in predator population = number of predators × growth of predator population/predator (5.14)

The components can be examined further. The growth of prey in the absence of the predator can be generalised as some function $f(H)$. The function may take specific forms such as exponential or logistic growth, or a non-specific form such as a function of rainfall or temperature, as described above.

The last term in equation 5.13 can be expanded to give:

$$\text{number of prey eaten by predators} = \text{number of predators} \times \text{number eaten/predator} \quad (5.15)$$

The number of prey eaten per predator is the per capita functional response (Holling, 1959; Krebs, 1985; Begon *et al.*, 1990). The per capita functional response can also be thought of as a general function, $g(H)$, which can take many specific forms. May (1981b) lists seven forms, from a constant to linear to curvilinear, and Eberhardt (1988) describes three forms, from linear to curvilinear. The functional response will be described further in sections 5.4 and 6.3.

The growth of the predator population, per predator, is the per capita numerical response. The per capita numerical response can also be thought of as a general function, $h(H)$, which takes several specific forms. May (1981b) lists three such examples which correspond to linear and curvilinear numerical responses. The relationships can be summarised as the general functions:

$$\text{change in prey population} = f(H) - P\,g(H) \quad (5.16)$$

$$\text{change in predator populations} = P\,h(H) \quad (5.17)$$

where the number of predators is P. May (1981b) described the predicted patterns of growth of such predator populations as cycles or oscillations converging to equilibrium. Caughley (1976) described one such model for an ungulate population. The interactive model estimated the maximum sustained yield occurred at about 70% of carrying capacity ($0.7K$), compared to the logistic model where it is at $0.5K$ (Fig. 5.3). Hence the predictions were closer to those of the generalised logistic model (Fig. 5.3). Crawley (1983) described the predicted dynamics of herbivory for many variations on the basic model.

A multi-trophic model of dynamics of kangaroo populations was described by Bayliss (1987) and Caughley (1987). The trends in kangaroo populations were linked to changes in food supply, which in turn was related to prior rainfall, with feedback effects incorporated. The model was an elaboration of the laissez-faire model described for herbivores (Caughley & Lawton, 1981) by adding vegetation growth as a function of prior rainfall and standing biomass of vegetation, instead of solely biomass. A graphical model of wolf dynamics linking wolves, moose and vegetation was described by Messier & Crete (1985). The moose population may be regulated at low densities by wolves or at high density by food shortage. The model was an elaboration of an earlier similar graphical model of Bergerud, Wyett & Snider (1983).

A different set of models have attempted to link animal abundance, rather than the change in animal abundance, to animal or habitat characteristics. For example, Damuth (1981) reported a power function between population density of mammals (D) and mean body weight (W). The relationship was:

$$D = a\ W^b \tag{5.18}$$

where a and b were constants estimated by regression. The regression accounted for 74% of variation in density. The mammals were mainly herbivores. Freeland (1990) has examined the relationship further for herbivores and separated the data for mammals within their natural distribution and mammals introduced to new environments. The latter in northern Australia consistently have population densities above that predicted by the density–body-weight equation. The higher densities were thought to be associated with fewer predators and parasites compared with the natural habitat of the mammals.

Braithwaite, Turner & Kelly (1984) reported relationships between soil parent material, foliage nutrients and abundance of arboreal mammals in south-eastern Australia. The relationships were not tested with independent data. The mammals were basically herbivores and were not vertebrate pests but the approach may be of relevance to pests.

A model for predicting the density of family groups of foxes in urban areas in Britain was described by Harris & Rayner (1986). The model used log-linear multiple regression to relate habitat features to density. Fox density increased with the amount of urban fringe and decreased with industry, fields, rented housing and housing occupation density. The predictions had to satisfy six prior criteria for the regression to be acceptable. The criteria were: (i) that at least 60% of overall variation in density must be accounted for, which corresponds to a correlation coefficient (r) of 0.775 (hence $r^2 = 0.60$ = the coefficient of determination), (ii) at least 50% of the predicted values must be within 1.0 units of true density and 75% within 1.5 units, (iii) the structure of the model must be consistent between areas, (iv) the model should be equally reliable between cities, (v) outliers should be explainable, and (vi) residual or error estimates should be explainable. The model was tested using independent data from a different city. Harris & Rayner (1986) cautioned that the predictions may be biased on urban fringes, in unusual habitats such as botanical gardens and where fox densities were very high or very low.

Thornton (1988) described a model for predicting distribution and density of badgers in south-western England. The model used multiple regression to relate density of main setts (burrows) to habitat characteristics. The fitted model had to satisfy two prior criteria, the first two of the six used by Harris & Rayner (1986), and was subsequently tested using independent data. The final fitted multiple regression related main sett density to five habitat variables: the 'diggability' of soils, topography, elevation, the length of hedgerows and the number of woodland units greater than 1 ha. The equation was used to predict sett density in 24 areas. The predicted values satisfied the second criterion, though not as well as the original data. An alternative criterion could have been used in both studies; the difference between the observed and predicted density was not significantly different from zero.

Further details of models or theories of population dynamics are described in Krebs (1985) and Begon *et al.* (1990). Both feature easy-to-read discussions of qualitative and quantitative aspects of dynamics. In later sections of this chapter and Chapter 6, a variety of the models described above will be used in models of damage and control.

5.3.2 *Dynamics of very small populations*

When a pest population has very few individuals the models described above may not be appropriate. Different models have been used to estimate the minimum viable population size and the probability and time to extinction. This is the pest control version of endangered species management, except here the objective may be eradication or control rather than conservation, and so the factors that increase the probability of extinction or decrease the time to extinction are of more interest.

In a small population the chance effects of events such as storms, droughts, demographic factors such as age and sex structure or effects of inbreeding may be more important than in a large population. As an example, Ralls, Brubber & Ballou (1979) reported that juvenile mortality of 15 species of inbred captive ungulate was higher than that of noninbred species. A large literature has developed around this topic (Soulé, 1987a). In this section a brief overview of the literature is given and readers referred to relevant studies.

The probability of extinction (p(Ext)) of a population was described by Krebs (1985) as:

$$p(\text{Ext}) = (d/b)^N \tag{5.19}$$

where d and b are the instantaneous death and birth rates respectively and N is the initial number of animals in the population. The equation is valid for $d < b$. An example for a population of feral pigs in Australia illustrates the equation. If the mortality rate is 0.00089/day and the birth rate is 0.0025/day, as estimated by Pech & Hone (1988), then as N increases the probability of extinction decreases. When $N = 2$, $p(\text{Ext}) = 0.127$ and when $N = 4$, $p(\text{Ext}) = 0.016$.

The population should increase exponentially as discussed by Krebs (1985) but may by chance go extinct, especially when N is very low. The equation is of relevance to control of isolated populations such as vertebrate pests on small islands or in fenced enclosures. An example of the latter was described for control of feral pigs in Hawaii (Hone & Stone, 1989). A management framework for use of such probability estimates is decision theory as described in section 4.2.4. Maguire (1986) and Maguire *et al.* (1988) described a decision analysis for management of endangered wildlife species.

The question of 'how many animals is enough' has been debated at length. Soulé (1987b) considered there is no universal or magic number that can be used in such debate and Shafer (1990) agreed. Each situation had to be assessed separately. Empirical data on mammals in several reserves in north-western America indicate that for those populations to survive an average of 75 years, initial median population sizes needed to be from at least 100 to up to several thousand (Soulé, 1987c). Berger (1990) reported that some isolated populations of bighorn sheep (*Ovis canadensis*) in south-western U.S.A. went extinct naturally. All populations of this species, which is not a vertebrate pest, that had fewer than 50 individuals went extinct in less than 50 years.

Harris & Allendorf (1989) reviewed nine estimators of the effective populations size (N_e) of a simulated grizzly bear (*Ursus arctos*) population. The effective population size is defined as the size of an idealized population that undergoes the same amount of loss of genetic variability as the real population under consideration (Koenig, 1988) or the number of individuals that contribute to future generations (Goodloe *et al.*, 1991). Begon *et al.* (1990) defined N_e as the size of a genetically idealized population to which the actual population is equivalent in genetic terms. Hence N_e is usually less than N, the actual population size. Harris & Allendorf (1989) did not clearly define N_e, but their review of estimators showed that many were biased, required data beyond most field studies, or had wide confidence intervals that negated clear interpretation of estimates.

The most accurate estimators of N_e were those of Hill (1972), Ryman *et al.* (1981) and a modification of that of Reed, Doerr & Walters (1986). For full details see Harris & Allendorf (1989). Koenig (1988) added an extra caution. Fluctuations in population size due to environmental changes may yield underestimates of the viable population size. The latter was defined by Shaffer (1981) as the smallest isolated population having a 99% chance of surviving for 1000 years. Goodloe *et al.* (1991) used the modified estimator of Reed *et al.* (1986) for three island populations of feral horses in the eastern U.S.A. The minimum effective population sizes ranged from 72 to 155 horses. Some such populations exceeded the estimated number needed to prevent excessive damage to native ecosystems.

The models of effective population size assume that the small population is isolated or closed. If immigration occurs then the effects of genetic inbreeding or loss of heterozygosity may be reduced. Maguire (1986) cited theoretical results that suggest an average of one unrelated immigrant per generation should be sufficient to overcome most deleterious effects of inbreeding in an otherwise closed population.

Models for estimating the duration of persistence of small isolated populations were reviewed by Burgman, Akcakaya & Loew (1988). They reported a wide variety of models, considerable disagreement in their predictions and the large amount of data needed to use the detailed models. Further research is needed here. Shaffer (1987) described predicted relationships between abundance and persistence of populations that were subject to one of three sources of uncertainty; catastrophic (such as floods, drought or fire), environmental (such as changes to food supply, parasites or predators) and demographic (changes in survival and reproduction). The predictions may have relevance to vertebrate pest control, particularly if control is viewed as a catastrophic source of uncertainty. When population size was high, catastrophic effects were the source of uncertainty acting the most strongly to reduce the time to extinction. When population size was very low the differences between the effects of uncertainty disappeared.

The generally theoretical models of small populations referred to above contrast with an empirical model described by Griffith *et al.* (1989). They reviewed the success of translocations of wildlife and factors influencing success. Translocations were defined as the intentional release of animals to the wild to establish, augment or reestablish a population. The success of the releases was estimated by logistic regression of the probability of survival and a string of environmental and population features. The features which were statistically significant were biological taxa (bird or mammal), the threatened status of the species, native game status, the habitat quality of the release area, location of release area, reproductive traits, the number of animals released and the duration of the program. The probability (p) of survival of a translocated population was given by Griffith *et al.* (1989) as:

$$p = 1/(1+e^{-x}) \tag{5.20}$$

where x is the sum of, the sum of coefficients and the values of categorical variables, and the sum of products of coefficients and the value of continuous variables.

The model may have application to vertebrate pest control through estimating the likelihood of survival of a colonising pest population. An example of the calculations is as follows. The numbers in brackets are the values of terms to be summed to estimate x based on coefficients in Table 2 of Griffith *et al.* (1989). Consider the case of a population of two (value $= \log_{10} 2$) wild boar (mammals $= 0.919$) introduced in one year (0.181×1) to an island of excellent habitat (1.681), within the core of the distribution of the species (1.028) as native game (0.972) and if wild boar are classified as early breeders (1.080). Wild boar can be pests of agriculture within their natural distribution (Mackin, 1970; Andrzejewski & Jezierski, 1978). The sum term (x) is equal to 6.162, so the probability of survival is:

$$p = 1/(1 + e^{-6.162}) = 0.998 \qquad (5.21)$$

Obviously, the estimated probability of survival is very high. The estimated probability of extinction is then 0.002, which is lower than that (0.127) estimated by equation 5.19. Griffith *et al.* (1989) showed that as the number of animals released increased the probability of survival of the population increased in a logarithmic manner, so the probability increased but at a declining rate. The results support the earlier work of Newsome & Noble (1986) suggesting that success at colonising is not a simple function of one ecological characteristic. Newsome & Noble (1986) considered that factors external to the coloniser, such as habitat characteristics, were most important. The general topic of modelling the dynamics of very small populations has still to find practical application to vertebrate pest control. However, the estimation of the probability of extinction and the time to extinction is potentially useful, though it often requires considerable data, not all of which is readily available. None of the available models explicitly incorporates reduction of abundance by control, though depending on the administration of the control, it may be equivalent to the effects of another source of environmental variability or a catastrophic source of variability.

5.3.3 *Movement*

Models of movement are concerned with describing patterns or rates of spread of pests and their impacts and describing what influences that movement or spread. Information on spread can then be used for strategic planning of how to restrict movement. Aspects of the strategic use were described in section 3.3. A good overview of biological invasions was given by Hengeveld (1989) and models and methods of parameter estimation described by van den Bosch, Hengeveld & Metz (1992).

Movement here includes all types of change in location by wildlife, including dispersal, movement within a home range or territory, and migration. Caughley (1980) described two models of dispersal – random and pressure. Random dispersal is the movement away from a birthplace to a place where an individual reproduces with the direction, timing and distance of movement being random. In pressure dispersal an individual leaves its birthplace when population density reaches a critical threshold. These two models of dispersal were regarded as extremes of a continuum of types of dispersal. Random dispersal generated a linear relationship between the radius of the area of distribution and time (Fig. 5.6). Pressure dispersal generated an exponential relationship (Fig. 5.6). Williamson & Brown (1986) considered that the spread of muskrat (*Ondatra zibethicus*) in Europe and grey squirrels (*Sciurus carolinensis*) in Britain were similar and supported an hypothesis of random dispersal, though in the case of the grey squirrel there were occasional periods of greater spread. The presence of a similar animal in the environment may alter the predicted rate of spread. Okubo *et al.* (1989) reported that the spread of the grey squirrel in Britain may have been slowed by the effects of competition from native red squirrels (*Sciurus vulgaris*), so the modelling of the spread as simple random dispersal may be simplistic. Hengeveld (1989) considered that the muskrat population in the Netherlands was stable at a low level until 1961. In that year the working week was shortened, presumably control efforts decreased, and subsequently muskrat spread through most of the country.

Caughley (1970a) and Hengeveld (1989) concluded that the pattern of spread of Himalayan thar (*Hemitragus jemlahicus*) after introduction into New Zealand was similar to that predicted for random dispersal though may be a bit more complex. Further data, and some recalculations, led Parkes & Tustin (1985) to conclude that the pattern of spread was consistent with the threshold pressure model of dispersal (Fig. 5.6).

Such observed patterns of spread are the end product of arrival, establishment and persistance. Separating the three components was discussed by Mollison (1986). Arrival is the process of getting from point a to point b. Establishment is the short-term increase in distribution and abundance that can occur after arrival. In deterministic models, establishment occurs when the

Fig. 5.6. The relationships between the radius of the area of distribution of a species that is spreading and the time since the start of such spread. The relationship for a population spreading by random dispersal is shown as a solid line and that of a population spreading by pressure dispersal as a dashed line.

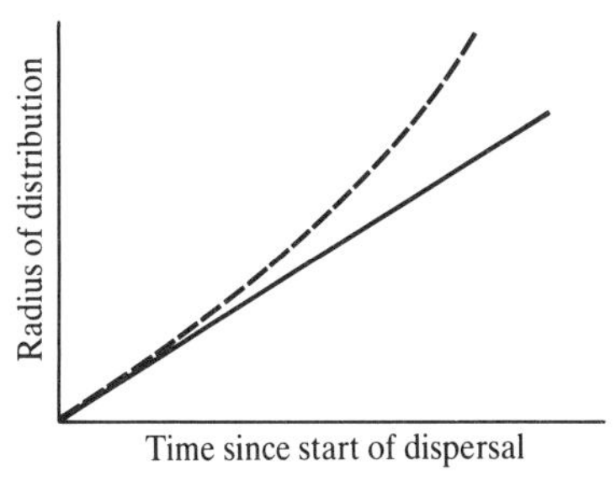

number of offspring per individual coloniser is equal to or greater than one. That is, each individual replaces itself. Persistance is the long-term survival of the spreading population.

The patterns of dispersal of foxes (Trewhella, Harris & McAllister, 1988) and factors affecting such dispersal in an urban fox population in Britain (Harris & Trewhella, 1988) have been described in detail, on the assumption that the rate of spread of rabies in foxes is influenced by fox dispersal. That assumption appears sound although the rate of rabies spread may be different to the rate of dispersal. Using these data the predicted patterns of fox dispersal in British urban areas have been described (Trewhella & Harris, 1988). The model simulated the first year of life of urban foxes. Spatial features of the model were based on a 500 m × 500 m grid, with foxes moving between cells in a grid. Temporal features were described by finite difference equations of changes in fox numbers due to mortality and dispersal. The direction of dispersal was modelled as a random variable and dispersal decreased in a negative exponential manner with increasing distance.

The spatial spread of infectious diseases such as rabies in European foxes has been carefully mapped (Macdonald, 1980) and modelled in studies including those by Kallen *et al.* (1985), Hengeveld (1989), van den Bosch, Metz & Diekmann (1990) and Murray & Seward (1992). The spread of an infectious disease is essentially an invasion of a host population by a pathogen, where the host is analogous to the food (Mollison, 1986).

The spatial spread of disease could be modelled in several ways: by diffusion, point-to-point, and grid-based models. Diffusion modelling usually assumes that disease spread occurs in one direction and can be characterised as a wave, though it may take the form of a plume of pathogens, especially when spread by wind (Gloster *et al.*, 1981; Donaldson, Lee & Shimshony, 1988). Point-to-point spread assumes that disease can move between points that are not necessarily adjacent, such as the spread of influenza between cities (Bailey, 1975). In grid-based models disease spread occurs between points in a matrix (Mollison, 1986). Grid-based models of rabies in urban foxes and tuberculosis in brushtail possums were reported by Smith & Harris (1991) and Barlow (1993) respectively.

Kallen *et al.* (1985) used a diffusion model to predict the pattern of spread of rabies in Britain. The spread was not uniform in all directions as the rate of spread was assumed to be related to fox population density, which was not homogeneous across Britain. The velocity of spread (c) of rabies in a fox population was estimated as:

$$c = 2\,(D\,(\beta\, s_0 - \alpha))^{0.5} \qquad (5.22)$$

where D is a diffusion coefficient, β is the transmission coefficient, s_0 is the initial density of susceptible foxes and α is the mortality rate of foxes caused by rabies. This equation predicts that the rate of spread (in km/y) increases according to the square root of initial fox density, the transmission coefficient, the mortality rate and the diffusion coefficient, which was estimated from:

$$D = k\ A \tag{5.23}$$

where k is the rate infective foxes leave their territory, so $1/k$ is the average time until a fox leaves its territory, and A is the area of the average territory. Murray *et al.* (1986) and Pech & McIlroy (1990) describe alternative methods for estimating the diffusion coefficient (D). The equation for the rate of spread can also be formulated (Kallen *et al.*, 1985; Dobson & May, 1986) as:

$$c = 2\ (D\ \alpha\ (R-1))^{0.5} \tag{5.24}$$

where R is the basic reproductive rate of the disease.

The spatial model of van den Bosch *et al.* (1990) predicts an increase then a slow decrease in rabies spread at high fox population densities. van den Bosch *et al.* (1990) incorrectly stated that the model of Kallen *et al.* (1985) predicted a linear relationship. Mollison (1991) reviewed both models and provides a more detailed comparison of their assumptions and predictions. Further data are needed to test these differing predictions. The predictions of both models are seemingly at odds with one set of empirical data on rabies spread in Germany. Macdonald (1980) cited data indicating that the average monthly rate of spread of rabies was independent of fox abundance.

Diffusion models of the spatial spread of rabies in foxes (Murray *et al.*, 1986) and foot and mouth disease in feral pigs (Pech & McIlroy, 1990) predicted that the rate of spread would increase in a curvilinear manner with increasing host carrying capacity (no./km^2), but not with susceptible host density.

In none of the models is the rate of spread of a disease predicted to be directly proportional to host density or abundance. Hence, reductions in host abundance will not always produce the same reduction in the rate of disease spread. This could be important in other situations of disease spread. For example, fences have been used in Zimbabwe to stop the spread of trypanosomiasis and foot and mouth disease (Taylor & Martin, 1987). It was considered by Taylor & Martin (1987) that the rate of spread of diseases was proportional to the number of animals that cross a fence, although no empirical data were presented to support the statement. The relationship could be considered linear for a small range of host abundance near zero, but the level of abundance at which the relationship becomes markedly curvilinear is not known. There is obviously a need to determine the range in host abundance over which linearity occurs, especially for foot and mouth disease, which is directly transmitted between hosts. As trypanosomiasis is spread by an insect vector the relationship will not be so simple.

5.4 Modelling damage

The literature on pests and their damage is reviewed briefly and then synthesised into a new model of the factors determining pest damage. As a general comment, the issue has not been explored in vertebrate pest research as extensively as in weed research. For example, see Doyle (1991) for models of crop yield and weed density relationships.

It is usually assumed that reducing pest abundance has benefits through reducing damage (Murton, 1968). After examining the functional response in predator–prey theory, Murton (1968) concluded that the amount of damage to a crop is not a simple function of pest abundance or crop density. The amount is related in a non-linear manner to food intake, which is a function of social behaviour, food availability and pest abundance. As a result, a unit change in pest abundance will not always give the same change in damage.

Cherrett *et al.* (1971) suggested that the extent of pest damage was a function of four variables: (i) the destructive potential per pest which may vary with pest age, size, genotype and environment, (ii) the duration of exposure, (iii) the resistance of the host or object being attacked, and (iv) the number of pests, so that the greater the number of pests, the greater the damage.

If pest damage is related to the above four variables, Cherrett *et al.* (1971) considered a simple relationship between pest abundance and damage will rarely occur. A linear relationship was considered less likely than a curvilinear relationship. The latter would occur as the first pests into, for example, a crop, occupy the best sites and subsequent pests occupy sites with lower yield and hence do less damage. Low pest abundance may cause limited damage that can be compensated for by, for example, increased plant growth. Only if pest abundance exceeds a threshold, where compensation does not exceed damage, does damage become apparent. Cherrett *et al.* (1971) noted that very low pest abundance may even increase yield, presumably because compensation exceeded damage. No examples from vertebrate pest control were cited, however.

Southwood & Norton (1973) suggested that a linear relationship is common between pest abundance and damage, at least until intraspecific competition occurred at high pest densities. In contrast, Woods (1974) considered that there was rarely a linear relationship between the number of pests and their damage. Norton (1976) described a model of pest damage that assumed a linear relationship between pest damage and pest abundance. The model was outlined in section 4.2.2.

That we can measure and predict damage is based on a simple theory of damage. This is a variation on a theory of response described by Dillon (1977), and is a set of assumptions describing inputs and outputs in a biological system. The theory assumes, firstly, that there is a continuous smooth causal relationship between the inputs (for example, pests, environment) and the outputs (the level of whatever is damaged). The relationship is not stepped or discontinuous. The second assumption is that the relationship may be linear or non-linear. In the broadest interpretation, the linear relationship can be regarded as a segment of the non-linear relationship.

Headley (1972a) assumed that pest damage (D) increased with increasing pest abundance (P):

$$D = a\,P^2 - b \tag{5.25}$$

where a is a constant and b is the damage threshold (Fig. 5.7a). The increasing damage function could be caused by direct facilitation between pests, which increases the damage per pest as pest abundance increases. Increasing food intake per pest would also generate such a relationship.
Product yield (Y) was assumed to be negatively related to damage:

$$Y = c - d\,D \tag{5.26}$$

where c is the yield when no damage occurs and d is the incremental yield effect (Fig. 5.7b). Yield is zero when $D = c/d$.

Fig. 5.7. Predicted relationships between: (*a*) Pest damage and pest abundance, as described by equation 5.25. The threshold level of pest abundance at which damage starts to occur is *b*. (*b*) Product yield and pest damage, as described by equation 5.26. Product yield is a maximum; *c*, when pest damage is zero and when pest damage is at a maximum, product yield is zero, *c*/*d*. (*c*) Product yield and pest abundance as described by equation 5.27. The numerical values of *a*, *b*, *c* and *d* in (*c*) are the same as in (*a*) and (*b*). (After Headley, 1972a.)

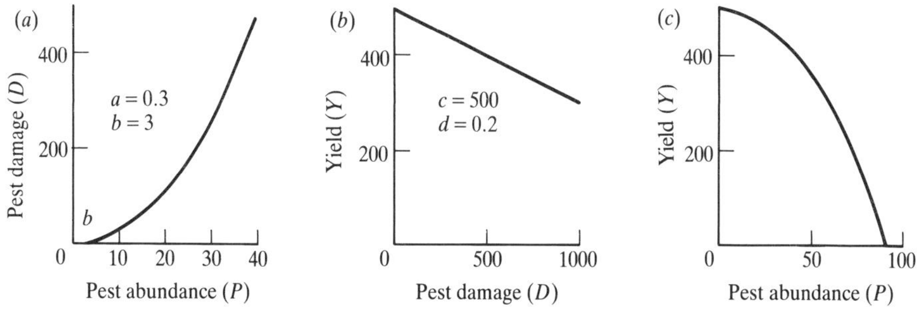

A relationship between pest abundance and product yield can be obtained by combining these two equations, as illustrated in Fig. 5.7c.

$$Y=(c+d\ b)-a\ d\ P^2 \tag{5.27}$$

The threshold pest abundance is estimated by $(c+d\ b)$, and the destructiveness of pests by the regression slope $(-a\ d)$. Yield is zero when $P=((c+d\ b)/a\ d)^{0.5}$. Tisdell (1982) criticised Headley's model as it did not consider mobility or fecundity of pests. Southwood & Norton (1973) considered that the relationship between pest abundance and yield was more complex. If pests attack the harvested product, for example, seeds or fruit, then increased pest abundance causes a rapid decline in yield. When pests attack the non-harvested product, for example leaves of a corn crop, yield is predicted to decline less rapidly (Fig. 5.8).

These two relationships can be viewed as one relationship with the curve for harvested products representing a continuation of the right-hand side of the curve for the non-harvested products (Conway, 1981). An ecological interpretation of pest damage – yield relationships was described by Hughes & McKinlay (1988). They showed by mathematical analysis that the pest abundance – damage relationship became non-linear with increasing aggregation (clumping) of the pests (Fig. 5.9a). Similarly, increased aggregation by pests generated a negative exponential yield – pest relationship (Fig. 5.9b). The suggested pest – yield relationship is similar to that described by Southwood & Norton (1973) but has a different explanation. Summers (1990) reported significant negative relationships between yield of wheat and an index of brent geese (*Branta bernicla*) abundance in eastern England. Such a linear relationship is intermediate between the relationships shown in Fig. 5.9b.

Fig. 5.8. A predicted relationship between product yield and pest abundance for harvested and non-harvested products. (After Southwood & Norton, 1973.)

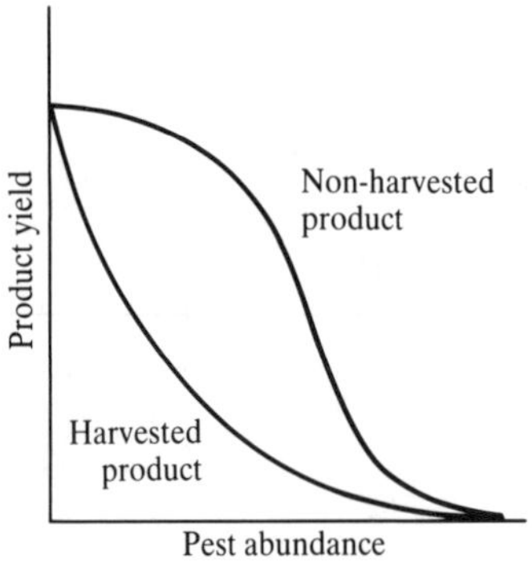

Crawley (1983) described models of herbivore attack on plants that had linear or non-linear damage functions, and which incorporated the spatial pattern of damage. The latter could be distributed according to a probability distribution such as the binomial, Poisson or negative binomial. Such models have apparently not been applied to vertebrate pest damage but could be.

A general model of vertebrate pest damage is developed here from the above literature and the functional response. Many cases of damage by vertebrate pests are a consequence of feeding behaviour of the pests. More damage is caused by individual pests eating more food. However, animals become satiated when large amounts of food are available. The functional response relationship in predator–prey and plant–herbivore theory (Holling, 1959, Murton, 1968; Starfield & Bleloch, 1986) describes the association between the amount of food available and the amount of food eaten (Fig. 5.10). Increasing

Fig. 5.9. Predicted relationships between (*a*) pest damage and pest abundance, and (*b*) product yield and pest abundance, for different degrees of pest aggregation. (After Hughes & McKinlay, 1988.)

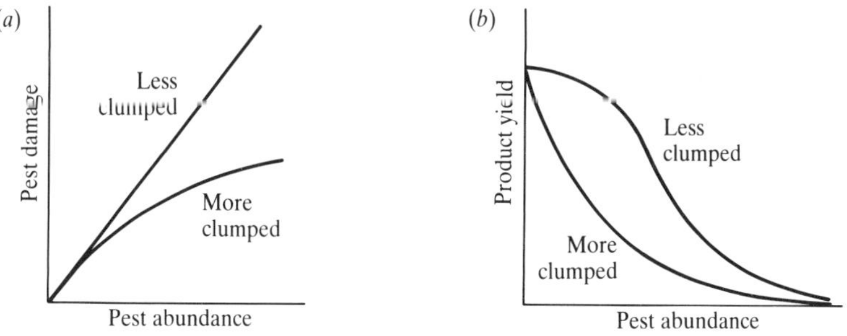

Fig. 5.10. Three functional response relationships, types I, II and III as described in the text, between forager intake and availability of food.

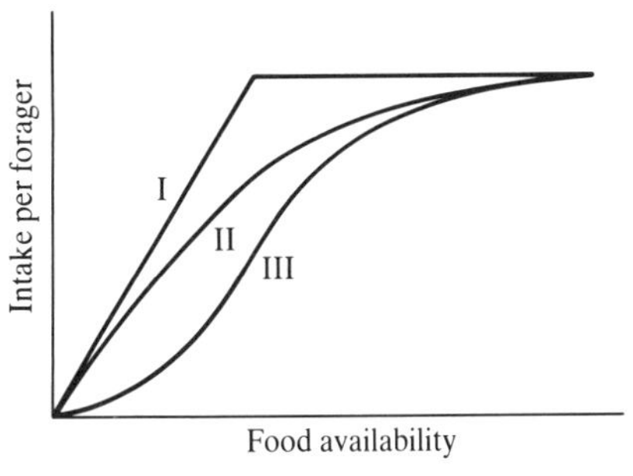

food availability results in increasing food intake, in either a ramp (I), a monotonically increasing (II) or a sigmoidal (III) relationship. Type II and type III relationships have been reported for vertebrates (Holling, 1959; Real, 1979; Short, 1985; Begon *et al.*, 1990). The type II relationship corresponds to a pest–damage function described by Hughes & McKinlay (1988) for an aggregated pest. Pech *et al.* (1992) reported a type III functional response for foxes (*Vulpes vulpes*) eating rabbits in semi-arid Australia. The report is a rare example of a study where type II and III responses were fitted and the curve of best fit (type III) estimated.

The literature relating pest abundance to damage was summarised in Table 2.1 for a wide range of pests and damage situations. Most (13/21) are statistically significant for tests of a linear relationship. The lack of significance in the other studies may be associated with curved functional responses, expected non-linear trends as for tuberculosis (discussed in section 5.6), density-dependence or features of a crop such as the area sown.

The first assumption of a general model is that the damage, per unit time, by pests (D) is proportional to the product of pest abundance $f(P)$ and the damage per pest (the functional response) ($g(V)$). This defines a general equation:

$$D = f(P)\, g(V) \tag{5.28}$$

A specific version of that general equation follows if the effects of pests is linear ($f(P) = bP$) and the functional response is assumed to be type II ($g(V) = 1 - e^{-dV}$). The specific equation is then:

$$D = b\, P\, (1 - e^{-dV}) \tag{5.29}$$

where V is the amount of product, for example crop, that is available for eating, b is the maximum rate at which a pest causes damage and d is the efficiency of feeding. This equation describes the patterns of pest damage shown in Fig. 5.11a,b. When the amount of product (V) is very large, the estimated damage saturates at the value bP. Equation 5.29 is of the same form as the fitted regression in Fig. 2.5 relating lamb kills by feral pigs to abundance of lambs. The term a (=26.50) in Fig. 2.5 has here been separated into two components, b and P. Since there were 15 adult feral pigs at the site (Pavlov *et al.*, 1981) $P = 15$, so $b = 26.50/15 = 1.77$. Equation 5.29 incorporates the significant relationships between damage and abundance reported in Table 2.1. The equation is an expanded non-monetary version of the model of Norton (1976) described in section 4.2.2.

A threshold (h) can be incorporated into the equation, corresponding to pests being present but no damage occurring.

$$D = -h + b\, P(1 - e^{-dV}) \tag{5.30}$$

The damage threshold ($D=0$) occurs when $P=h/b(1-e^{-dV})$ or $V=-1/d(\ln(1-h/b\,P))$. The threshold can be graphed as shown in Fig. 5.11c.

The relationship between inputs (pest abundance and product availability) may be different when more than one pest is present because of interference or social facilitation. This possibility can be crudely modelled by the inclusion of an exponent (c) on the term for pest abundance. When only one pest animal is present $P=1$ and $c=0$. When interference occurs between pests then $0<c<1$, and when social facilitation occurs $c>1$. This is shown in the following equation and the various combinations are shown in Fig. 5.12a,b,c.

$$D=-h+b\,P^{c}\,(1-e^{-dV}) \tag{5.31}$$

Headley's (1972a) model, equation 5.25, occurs when $c=2$ and V or d is large. The various parameters in the above equation can be estimated by non-linear least squares of experimental results when P, V and D are known, or maybe by least squares regression similar to the analysis of numerical responses by Bayliss (1987). A comparative study is needed of methods for estimating parameters of the numerical response. These relationships will be appropriate only if there is no compensatory growth and change in yield by a crop. If compensatory growth occurs then the relationships describe net damage not total damage. Similarly, the model assumes that there are no time lags between the actions of the pests and the effects.

The model of blackbird damage to crops (Wiens & Dyer, 1975) did not report the predicted relationship between bird abundance and damage. The model was based on energy requirements of the birds and that was estimated to increase at a decreasing rate as a power function of body weight. The model

Fig. 5.11. Predicted relationships between pest damage per unit time and (*a*) pest abundance, (*b*) product availability with no threshold, as estimated by equation 5.29, and (*c*) product availability with a threshold, as estimated by equation 5.30.

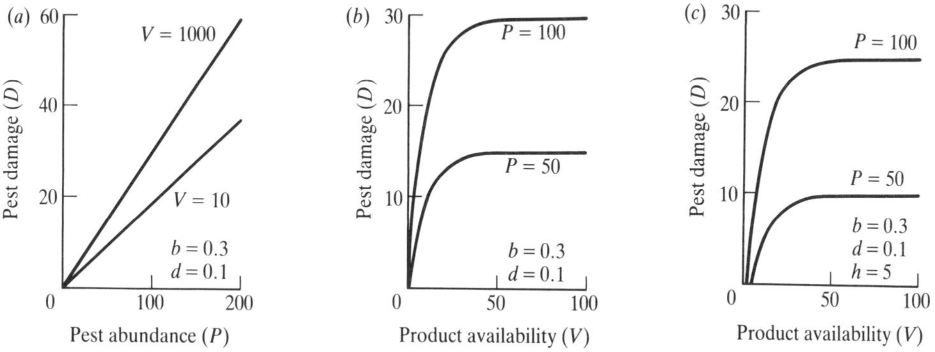

would then predict a curvilinear relationship between bird abundance and impacts such that $0<c<1$. Tisdell, Auld & Menz (1984b) considered that weeds may cause yield loss in crops or pastures. They suggested three forms of the relationship between weed density and yield loss that are the three forms shown in Figure 5.12. Izac & O'Brien (1991) suggested that the relationship between abundance of feral pigs and the level of damage was curvilinear. Their relationship corresponds to the curve in Figure 5.12a when $0<c<1$.

The model suggests that the response to pest control of, for example, damaged vegetation, may not always be linear. A unit reduction in pest abundance may produce variable results in damage depending on the pest–damage relationship. Nugent (1990) suggested that reduction of deer (*Dama dama*) in Otago, New Zealand, may produce variable responses in forest vegetation because of the differing palatability of leaves. If a diverse food supply is available to a pest then a type III functional response is probably most appropriate in the general model. Such a response would generate a sigmoidal relationship between damage and abundance and a curvilinear response to deer control.

The above model may explain why some studies, as described in Table 2.1, have reported a significant correlation between pest abundance and the level of damage and other studies have reported no significant correlation. The underlying relationship may be linear or non-linear. If only a linear hypothesis

Fig. 5.12. Predicted relationships between pest damage per unit time and (*a*) pest abundance, (*b*) product availability with no damage threshold and (*c*) product availability with a damage threshold, as estimated by equation 5.31. The effects of varying levels of pest interference or facilitation (parameter *c*) and efficiency of feeding (parameter *d*) are also shown.

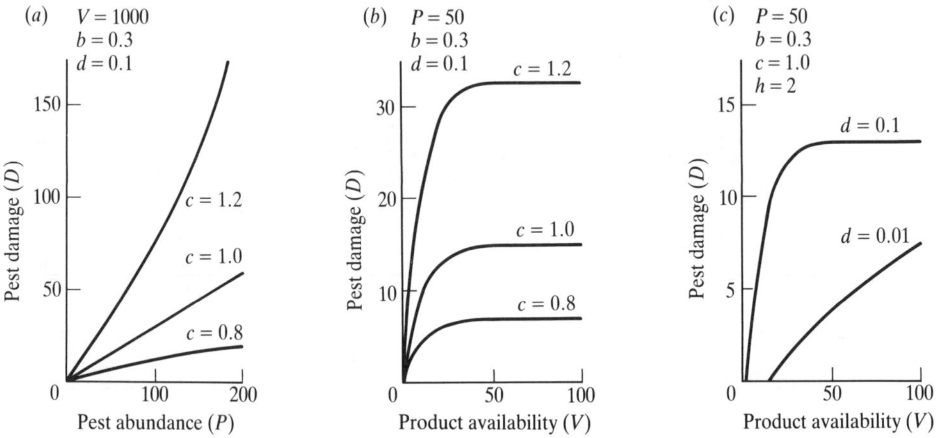

is tested then non-linear relationships will not be detected. Also the relative effects of abundance (P) and food (V) may vary between studies.

The model suggests that various actions could be taken to reduce damage by vertebrate pests (Table 5.1). The options are strategic as the parameters of the model have not been estimated, so the predictions cannot be exact. The trends shown in Figures 5.11 and 5.12 are testable hypotheses. There is a need for such testing.

5.5 Predation of livestock

Statistical analysis of predation was reviewed in section 2.4 and analysis of predation control in section 3.15. Economic analysis of predation control was reviewed in section 4.4.

Predation has been modelled many times in ecology, for example as discussed in May (1981b), Krebs (1985) and Begon *et al.* (1990). Much of the older modelling has focussed on the functional and numerical responses as components of modified Lotka–Volterra models, or on other aspects of foraging theory as discussed by Stephens & Krebs (1986). Given the obvious parallels between predation in ecology and predation of livestock it is surprising that the latter topic has not apparently been investigated mathematically. The topic is not explored in detail here, rather, readers are referred to the above references and to sections 5.4 and 6.5 of this book. Smith *et al.* (1986) used a difference equation model of annual dynamics of a coyote population to predict future trends in abundance. In the model natural mortality was estimated iteratively until the population trend approximated the known estimates of coyote abundance.

5.6 Infectious diseases

Statistical analysis of damage caused by infectious diseases was reviewed in section 2.5 and analysis of disease control in section 3.16.

Table 5.1. *Strategic options for reduction of damage by vertebrate pests, based on equation 5.31*

1. Increase damage threshold (h)
2. Decrease rate of damage (b)
3. Reduce the number of pests (P)
4. Change the behaviour of pests so that c decreases
5. Reduce the efficiency of feeding (d)
6. Reduce the amount of product available to be damaged (V) (unlikely to be popular!)

Economic analysis were reviewed in section 4.5. This section reviews some models of infectious diseases and examines a model of fox rabies as a case study.

Vertebrates can be pests because of their role as hosts for disease pathogens. Foxes as hosts of rabies, badgers as hosts of tuberculosis, and rodents as hosts of plague are well-known examples. These diseases have wild animals and humans as hosts and hence are of direct interest to public health programs. The dynamics of diseases in host populations has been studied intensively by mathematical modelling (Bailey, 1975; Anderson & May, 1979; 1986, May & Anderson, 1979; Bradley, 1982; Anderson & May, 1991). The discussion by Bradley (1982) of the history and application of disease modelling is very useful and sobering. Some features and outcomes of three classes of model are described here: those involving one host species and one pathogen, those with one host species, one pathogen and one species as a vector of the pathogen, and, briefly, those involving more than one host species. The discussion will concentrate on deterministic models, that is, models that assume parameters to be constant, so that for one input there is only one output. The discussion is mostly about microparasites (Anderson & May, 1979) such as viruses and bacteria. Helminths are classified as macroparasites and the dynamics of their interactions with hosts has been modelled extensively elsewhere (May & Anderson, 1979).

The effects of an infectious disease on a host population have been modelled by describing the host population as consisting of a series of groups of individuals. For example, for a hypothetical disease all individuals start as susceptible to infection. Some get infected and, after an incubation period, start to show signs of the disease and become infectious. Then those individuals may die from the disease or become immune to it. That immunity may be lost over time. In modelling, the flow of individuals between these groups can be described by equations, as the flow occurs at certain rates. One approach, of deterministic modelling, is to describe such rates by differential equations (Anderson & May, 1979; May & Anderson, 1979). That is the approach used here. The structure of such a host–disease interaction is shown in Fig. 5.13.

The rate at which susceptible animals (X) acquire infection is proportional to the number of infectious animals (Y) times the number of susceptible animals. The proportionality constant is the transmission coefficient (β) and βXY is called the transmission term. This is the mass-action assumption of models of infectious diseases, and is based on the assumption of homogeneous mixing of infectious and susceptible hosts. The assumption is simplistic but is the standard starting point in modelling. It is probably as unrealistic as the analogous linear functional response in the original Lotka–Volterra model of predator–prey interactions. Variations on this assumption have been examined, theoretically, by May & Anderson (1979) using a saturation equation,

Liu, Levin & Iwasa (1986) and Liu, Hethcote & Levin (1987). The latter study showed that predictions are most sensitive to the estimated number of infectives (Y) and examined the more general transmission term $\beta X^p Y^q$, which reduces to the basic transmission term when $p = q = 1$.

The latent period is the mean time between contracting infection and the onset of pathogen excretion, and is equal to $1/\sigma$. Subsequently, animals recover at a rate v, so the length of the infectious period is $1/v$. Animals die from the infection at a rate α. Animals lose immunity at a rate γ, so the duration of the immune period is $1/\gamma$. The total number of hosts is simply the sum of the number in each class ($N = X + I + Y + Z$) where I is the number of latent hosts (infected but not infectious) and Z is the number of immune hosts. The units for X, I, Y and Z can be individuals or individuals/area, but all have to be one or the other. The rate of movement of animals through the groups can now be described by the following equations.

$$dX/dt = -\beta XY + \gamma Z \tag{5.32}$$

$$dI/dt = \beta XY - \sigma I \tag{5.33}$$

$$dY/dt = \sigma I - \alpha Y - vY \tag{5.34}$$

$$dZ/dt = vY - \gamma Z \tag{5.35}$$

The host population has no natural dynamics as the birth and natural death rates are assumed to be zero. If the host population is assumed to have a per

Fig. 5.13. Flow chart of the movement of animals through four population compartments during an outbreak of an infectious disease; susceptible, latent (infected but not infectious), infectious and immune. In this simplified model there are no births or natural deaths. See text for definition of symbols.

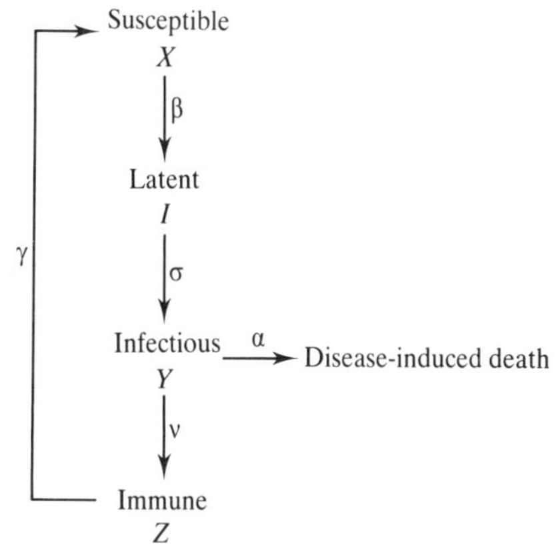

capita birth rate a, and a per capita rate of natural deaths, b, in the absence of the pathogen, then population growth is the difference between births and deaths $(a-b)$ and is therefore exponential. This is usually an unrealistic assumption but will illustrate the approach. It is assumed that births are density-independent and occur only to susceptible hosts but any hosts can die naturally irrespective of their disease status. The equations are now:

$$dX/dt = -\beta XY + \gamma Z + aN - bX \quad (5.36)$$

$$dI/dt = \beta XY - \sigma I - bI \quad (5.37)$$

$$dY/dt = \sigma I - \alpha Y - vY - bY \quad (5.38)$$

$$dZ/dt = vY - \gamma Z - bZ \quad (5.39)$$

If infected females give birth to infected young (vertical transmission) then equation 5.38 needs to include a term for such births. The modelling of population growth in the absence of the pathogen could be changed to be logistic growth by including a term for density-dependent mortality (g), as done by Anderson *et al.* (1981) for foxes (see equations 5.48 to 5.50), and by Pech & Hone (1988) and Pech & McIlroy (1990) for feral pigs, or changed to reflect environmental variability such as by using the numerical response. Hence, any relevant models of population dynamics described in section 5.3.1 could be inserted here.

A disease with dynamics like those encapsulated in equations (5.36) to (5.39) is expected to show a large pulse of infection shortly after introduction of the pathogen, followed by damped oscillations to equilibrium (Anderson, 1982b). This will occur if there are sufficient animals in the population, or, expressed differently, if the abundance of susceptibles is greater than the threshold abundance necessary for disease maintenance.

Trends in southern Australia in the number of rabbits with myxomatosis and the number of rabbit carcasses after deliberate introduction of the pathogen were of damped oscillations (Myers, 1954). In this case the oscillations were declining when observations ceased. It was considered that the spread of myxoma virus occurred directly between infectious and susceptible hosts and vectors were not involved or that the abundance of mosquitoes was below the threshold density for effective transmission. The spread of pathogens by vectors is a bit more complex than implied by Myers (1954) as it is determined by the ratio of vectors to hosts rather than simply the abundance of vectors; this is described later in this section. A series of damped oscillations in the number of dead wild rabbits in France was reported by Arthur & Louzis (1988) following introduction of myxomatosis in June 1952. Peaks in deaths were in 1953, 1957, 1960 and 1965. The data ended in 1967.

The host population size described by the above equations can be depressed by the infection. The pathogen may persist or disappear. The latter will occur if each infectious individual on average infects less than one susceptible animal before the infectious animal dies, for whatever reason. The number of cases of infection is the number of secondary cases per infectious individual during its infectious lifespan (Anderson, 1984). This is also known as the basic reproductive rate of the disease (R). Expressed differently, if $R<1$ the infection disappears from the host population. If $R \geqslant 1$ the infection establishes. The above model can be solved to estimate R and the related parameter the threshold host abundance. The latter is the number of susceptible animals in the host population needed to sustain the infection.

The threshold host abundance, or density, is estimated by solving the equations for the equilibrium situation. At equilibrium the rates of change of the number of susceptibles and so on (dX/dt, dI/dt ...) equal zero. The equations can be solved to estimate the threshold host density (K_T). For example:

$$dI/dt=0=\beta XY-\sigma I-bI \qquad (5.40)$$

$$dY/dt=0=\sigma I-\alpha Y-vY-bY \qquad (5.41)$$

Therefore, solving dI/dt for I gives:

$$I=\beta XY/(\sigma+b) \qquad (5.42)$$

Substituting for I in dY/dt gives:

$$0=\sigma\beta XY/(\sigma+b)-\alpha Y-vY-bY \qquad (5.43)$$

which can be rearranged, after cancelling Y, to give X as the left-hand side of the equation. The value of X is symbolised by K_{T}.

$$K_{\mathrm{T}}=(\sigma+b)\ (\alpha+v+b)/\beta\sigma \qquad (5.44)$$

The main implication of this result for vertebrate pest control is that disease control or disease eradication does not require host eradication. The only requirement is that the abundance or density of susceptible hosts is less than the threshold abundance or density. That of course may be easier said than done. Anderson & May (1979) and May & Hassell (1988) showed the effects of various characteristics of the host–pathogen interaction on the equation for the threshold host density. These are summarised in Table 5.2. Diseases which are sexually transmitted may have a very low threshold abundance as the spread is effectively independent of host population size (Anderson & May, 1979; Anderson, 1984).

If the susceptible population grows, in the absence of the pathogen, in response to food supply or rainfall as in the numerical response (Fig. 5.5), then that has an influence on the estimated threshold host density (K_T). That can be seen from the derivation above, which uses the natural mortality rate, which would be determined by such effects of food or rainfall.

The basic reproductive rate of the infection (R) is defined as the ratio of the number of susceptibles to the threshold density:

$$R = X/K_T \tag{5.45}$$

where R is a dimensionless quantity. Substituting for K_T in equation 5.45 from equation 5.44 gives:

$$R = \beta X\sigma/(\sigma + b)(\alpha + v + b) \tag{5.46}$$

If there is no incubation period then this reduces to:

$$R = \beta X/(\alpha + v + b) \tag{5.47}$$

The predicted number of secondary cases is directly proportional to the abundance of susceptible hosts. Mollison (1984) suggested, in contrast, that secondary infections should increase more slowly than abundance. The prediction of direct proportionality is most easily interpreted as that expected at the start of an epidemic when one infected host is introduced into a

Table 5.2. *The effects of various characteristics of the host–pathogen interaction on the estimates of the threshold host density (K_T) for directly transmitted infections*

Characteristic	Effect on K_T
Long incubation period	Increase
Short period of infection	Increase
High disease-induced mortality	Increase
High natural mortality	Increase
High transmission rate	Decrease
Vertical transmission (across the placenta) with reduced births by infected females	Decrease
Long-lived free-living infective stages	Decrease
Sexual transmission	Decrease towards zero
Transient or life-long immunity	No effect
Reduction of birth rate	No effect (if population growth exponential when disease absent)

After Anderson & May (1979), May & Hassell (1988).

population of susceptible hosts. If the number of susceptibles is greater than the threshold density then $R > 1$ and the disease will establish (become endemic). If the number of susceptibles is less, the disease disappears ($R < 1$) (Fig. 5.14). As the infection spreads then the number of susceptibles typically declines so the predicted number of secondary cases also declines. Equation 5.46 shows which aspects of the interaction between the host and pathogen determine the number of secondary infections. Typically, the value of R is most influenced by the value of the transmission coefficient (β) and the number of susceptible hosts (X).

Vertebrate pest and disease situations have been studied intensively using these types of model, especially rabies in foxes (Macdonald, 1980; Anderson *et al.*, 1981; Bacon, 1985; Smith, 1985) and raccoons (Coyne *et al.*, 1989), myxomatosis in rabbits (Anderson & May, 1982a; Dwyer *et al.*, 1990), tuberculosis in badgers (Anderson & Trewhella, 1985) and brushtail possums (Barlow, 1991a,b, 1993), foot and mouth disease in feral pigs (Pech & Hone, 1988; Pech & McIlroy, 1990), classical swine fever (Hone, Pech & Yip, 1992) and a liver parasite *Capillaria hepatica* of house mouse (McCallum & Singleton, 1989).

The models used in those studies differed slightly from the general model described in equations 5.36 to 5.39. In the fox models, for example, infected animals do not recover from rabies ($v = 0$), so there is no group of immune individuals. Tuberculosis may occur in badgers that show no clinical signs of the disease (carriers) or some badgers may be infected but are not infectious (inactive). Each of these events requires an alteration to the basic disease model. The tuberculosis (TB) model of Anderson & Trewhella (1985) was structured slightly differently, reflecting the possibility of badgers getting

Fig. 5.14. The effect of different values of the basic reproductive rate (R) of a disease on the predicted number of infectious hosts, with increasing time since introduction of one infected host into a population of susceptibles.

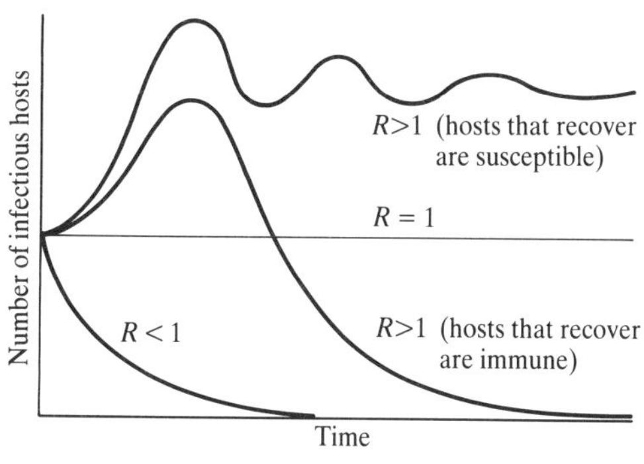

infected from infected hosts or from infective stages of the pathogen in the environment. The model of TB in possums assumed non-homogeneous mixing of susceptibles and infecteds (Barlow, 1991a,b, 1993) so had different equations for the basic reproductive rate. The equation for the threshold density of feral pigs (Pech & Hone, 1988) differed from equation (5.44) as it includes the birth rate (a) rather than the natural death rate (b). This is a consequence of assuming that population growth in the absence of the disease was logistic rather than exponential. The model of *Capillaria* in mice was of a macroparasite so included a description of the frequency distribution of parasites per host (McCallum & Singleton, 1989).

Estimates of parameters in the models were obtained from the literature or in the case of the transmission coefficient indirectly from field observations. Empirical estimates of the threshold density have been reported in several studies. Anderson *et al.* (1981) reported that the threshold population density of foxes for rabies was 1.0/km^2. Anderson & Trewhella (1985) estimated that the threshold population density of badgers for tuberculosis was 1–5/km^2 and Stuart & Wilesmith (1988) reported that the lowest threshold may be 5/km^2. Coyne *et al.* (1989) reported that the threshold population density of raccoons for rabies was 3.0/km^2. The method of estimating K_T from field data as used by Anderson *et al.* (1981), Stuart & Wilesmith (1988) and Coyne *et al.* (1989) may give biased results. The estimates use the population density of hosts where the disease does not currently occur. The population may not consist entirely of susceptible hosts but may also include immune hosts, though these should be very rare for fox rabies. The threshold density is, strictly, the density of susceptibles, so use of uncorrected field data may overestimate the threshold density. Hence, use of such estimates of K_T for estimating the transmission coefficient will tend to give underestimates, following equation 5.44.

Of all the parameters in any disease model, the transmission coefficient is usually the hardest to estimate. A list of the methods used is given in Table 5.3. The methods require very different data, some of which are usually not available, such as initial host abundance.

There have been few empirical tests of the predictions of the disease models. A prediction of the model of tuberculosis in badgers (Anderson & Trewhella, 1985) was that tuberculosis should depress badger abundance below that of disease-free populations. Cheeseman *et al.* (1988) reported preliminary data that did not support this prediction. There was no significant correlation between the number of badgers with tuberculosis and the number of badgers ($r = -0.12$, df$=17$, $P > 0.05$). It was not established, however, whether the badger population was above the threshold density, though Anderson & Trewhella (1985) suggested that K_T would be very low. Anderson & Trewhella

(1985) reported that there were insufficient data to reach a conclusion.

In contrast, Cheeseman *et al.* (1981) reported that there was no evidence of a relationship between group size and prevalence of TB but did not actually calculate a relationship such as a correlation coefficient, though Coleman (1988) cited the result as a negative correlation. My analysis of data in Table 1 of Cheeseman *et al.* (1981) showed a significant correlation ($r=0.58$, df$=22$,

Table 5.3. *Methods for estimating the transmission coefficient in compartment models of the dynamics of infectious diseases, including pathogens and parasites. Methods 1 to 8 are for deterministic models and 9 to 11 for stochastic models. Method 4 estimates the force of infection*

No.	Method	Disease or pathogen	Source
1.	Estimate by regression of prevalence and time since start of epidemic, when population size constant	*Pasteurella muris*	Anderson & May (1979)
2.	Estimate from equation for threshold density	Rabies	Anderson *et al.* (1981); Coyne *et al.* (1989)
3.	Estimate from equilibrium prevalence	—	May & Anderson (1983)
4.	Estimate from age-prevalence data	*Schistosoma japonicum*	Cohen (1973, 1976)
5.	Infer from behaviour data	Foot and mouth disease	Pech & Hone (1988)
6.	Infer from 'contacts' between hosts	Foot and mouth disease	Pech & McIlroy (1990)
7.	Estimate from number of secondary infections and other disease parameters	Insect virus	Dwyer (1991)
8.	Estimate by regression of trends in deaths and time since start of epidemic	Classical swine fever	Hone *et al.* (1992)
9.	Estimate from known initial host population size and total deaths in epidemic	Classical swine fever	Yip (1989); Hone *et al.* (1992)
10.	Estimate from known initial host population size and trends in disease prevalence	—	Becker (1977)
11.	Estimate from daily trends in number of susceptibles, infectives and host abundance	Myxomatosis	Saunders (1980)

$P<0.01$) between the number of badgers infected (y) and the number of badgers in each social group (x), and between the percentage of badgers infected (y) and the number of badgers in each social group (x) ($r=0.43$, df$=22$, $P<0.05$). Cheeseman *et al.* (1989) also presented data but did not analyse them. My analysis of data in their Table 1, shows a non-significant correlation ($r=0.43$, df$=14$, $P>0.05$) between mean number of social groups/km^2 (x) and prevalence (y). Data in their Table 2 showed another non-significant correlation ($r=0.11$, df$=4$, $P>0.05$) between population density (x) and prevalence (y). The underlying relationship may not always be linear because of the non-linearity of the transmission process. Anderson & Trewhella (1985) did not expect a linear relationship between host density and equilibrium prevalence. A highly curvilinear trend between host density and prevalence (not equilibrium but current) can be inferred from Fig. 13 in Anderson & Trewhella (1985).

A significant positive relationship between rabbit abundance and prevalence of myxomatosis in a rabbit population in southern England was reported by Trout *et al.* (1992). In contrast, Ross *et al.* (1989) reported a non-significant relationship when data were pooled across three sites.

Fox rabies is now described in more detail as a case study. Rabies is a disease caused by a virus. The disease occurs in Europe, Asia, the Americas and Africa (Bacon, 1985). The dynamics of the disease in foxes have been modelled extensively. I will describe one simplified version of a model, that originally reported by Anderson *et al.* (1981) and Anderson (1982c). The model is a deterministic compartment model with three classes of fox: susceptible (X), latent (incubating) (I) and infectious (Y) (as in Fig. 5.13 but without immune hosts). The model of Kallen *et al.* (1985) assumed there was no latent period, whereas a model by Murray *et al.* (1986) included a latent period. Foxes invariably die after being infected with rabies virus, so no animals recover or develop immunity.

The rate of movement of foxes through the classes was described by the following equations, assuming births (a) and natural deaths (b) occur. The total number of hosts is the sum of the number in each class ($N=X+I+Y$). The units for X, I and Y were foxes/km^2.

$$\mathrm{d}X/\mathrm{d}t=-\beta XY+aX-bX-gNX \qquad (5.48)$$

$$\mathrm{d}I/\mathrm{d}t=\beta XY-\sigma I-bI-gNI \qquad (5.49)$$

$$\mathrm{d}Y/\mathrm{d}t=\sigma I-\alpha Y-bY-gNY \qquad (5.50)$$

The parameters σ, α, a, b, and β were estimated from empirical data. The latent period was assumed to be 29 days, so the per capita rate of change of latent foxes (σ) was $(1/29) \times 365 = 13$/year. Infectious foxes were assumed to live for five days so the per capita rate of rabies-induced mortality (α) was $(1/5) \times 365 = 73$/year. The per capita birth rate (a) was estimated to be 1/year. The per capita natural death rate (b) was estimated to be 0.5/year and the intrinsic rate of increase of the population (r_m) was 0.5/year. It was assumed the fox population increased according to logistic growth in the absence of rabies, so the equations include a term for density-dependent mortality (gN) where carrying capacity $K = r_m/g$. In this model the density-dependence was assumed to occur for each class of fox, but in contrast, births were only to and of susceptible foxes. The transmission coefficient (β) was estimated from an equation for the threshold density (K_T) and the parameter estimates (method 2 in Table 5.3). The threshold density was assumed to be 1 fox/km^2 based on European experience. Given that the threshold density was:

$$K_T = (\sigma + a)(\alpha + a)/\beta\sigma \tag{5.51}$$

then:

$$1.0 = (13 + 1)(73 + 1)/13\beta \tag{5.52}$$

which solves to $\beta = 79.7$ km^2/fox/year. The units of β are different to those reported by Anderson *et al.* (1981). The change is necessary to balance units in equations 5.51 and 5.52, and follows a discussion by Caley (1993) of units for the transmission coefficient in such disease models. As a consequence, many authors have incorrectly cited the units of the transmission coefficient. Murray *et al.* (1986) indicated that lower values of the threshold density may also occur. Equation 5.51 is equation 5.44 except here foxes do not recover ($v = 0$) and b is replaced by $b + gN$ which at equilibrium equals a. The equations for the model are:

$$dX/dt = -79.7XY + 1X - (0.5 + gN)X \tag{5.53}$$

$$dI/dt = 79.7XY - (13 + 0.5 + gN)I \tag{5.54}$$

$$dY/dt = 13I - (73 + 0.5 + gN)Y \tag{5.55}$$

The density-dependent mortality term (g) has not been estimated here as it disappears from the equation when estimating the threshold fox density. The basic reproductive rate of the fox rabies is given by equation 5.46, after relevant modifications, so $R = \beta\sigma X/(\sigma + a)(\alpha + a)$ which is the inverse of equation 5.51 when $X = K_T$.

When the carrying capacity was 2 foxes/km^2 the model predicted that fox population density and rabies prevalence ($100Y/N$) would oscillate to an equilibrium. The period of the oscillations was about three to five years, similar to that observed in the field. The predictions of the model differed when the basic assumptions were varied. If the latency period ($1/\sigma$) was increased and carrying capacity (K) was high, then disease prevalence would cycle and not show damped oscillations.

The model is quite simplistic. It ignores alternative hosts, alternative patterns of population growth by rabies-free foxes, spatial spread and stochastic and demographic effects. Yet its simplicity makes it a good starting point for further work. More detailed models of fox rabies have been described and the sources of some are shown in Table 5.4. Smith & Harris (1991) have a useful summary (their Table 1) of the characteristics of most models.

Many pathogens are spread indirectly between hosts via vectors rather than spread directly. For example, myxomatosis can be spread between rabbits by mosquitoes or fleas and plague can be spread between hosts by infected fleas. Trypanosomiasis, a disease of humans, livestock and wild animals, is spread by the tsetse fly (*Glossinia* spp.) (Taylor & Martin, 1987). Alternative hosts are sometimes, and incorrectly, called disease vectors. Hosts are animals in which a pathogen can reproduce, but pathogen reproduction does not occur in a vector.

The different transmission process associated with vectors necessitates changes in the above models, and has implications for disease control. Mathematical models of the dynamics of myxomatosis in rabbits have been published, including those by Saunders (1980), Anderson & May (1982a) and Dwyer *et al.* (1990). Each assumed that disease transmission occurred directly between hosts rather than between hosts and vectors. The predictions of the models should be treated accordingly. May & Anderson (1979) and Anderson (1981, 1984) described models of diseases spread by vectors. The models were greatly simplified and are so here. Modelling the dynamics of a vector-spread disease involves extending the basic model by describing the changes in the vector and the host populations. Hence, a set of equations are derived for each. Now disease transmission occurs between infected vectors (Y') and susceptible

Table 5.4. *Sources of some models of the dynamics of rabies in fox populations*

Type of model	Non-spatial	Spatial
Deterministic	Anderson *et al.* (1981) Anderson (1982c) Smith (1985)	Kallen *et al.* (1985) Murray *et al.* (1986) van den Bosch *et al.* (1990)
Stochastic	Smart & Giles (1973)	Ball (1985), Mollison & Kuulasmaa (1985), Hengeveld (1989), Smith & Harris (1991)

hosts (X), and between infected hosts (Y) and susceptible vectors (X'). If the susceptible and infected vectors have the same biting rates and the vector makes a fixed number of bites per day, then there is only one transmission rate or coefficient (β'). New estimates are needed of the latent period and the duration of infectivity in the vector. Fig. 5.15 is a diagram of the model. The equations for the dynamics of the infection, with births and natural deaths included, though not shown in Fig. 5.15, are:

$$dX/dt = -\beta'(X/N)Y' + \gamma Z + aN - bX \quad (5.56)$$

$$dI/dt = \beta'(X/N)Y' - \sigma I - bI \quad (5.57)$$

$$dY/dt = \sigma I - \alpha Y - vY - bY \quad (5.58)$$

$$dZ/dt = vY - \gamma Z - bZ \quad (5.59)$$

$$dX'/dt = -\beta' X'(Y/N) + a'N' - b'X' \quad (5.60)$$

$$dI'/dt = \beta' X'(Y/N) - \sigma' I' - b'I' \quad (5.61)$$

$$dY'/dt = \sigma' I' - b'Y' \quad (5.62)$$

As before, the basic reproductive rate (R') is estimated from the equilibrium conditions. The derivation is not shown but gives:

$$R' = \sigma\sigma'\beta'^2\ (N'/N)/((\sigma+b)\ (\alpha+v+b)\ (\sigma'+b')b') \quad (5.63)$$

Fig. 5.15. Flow chart of the movement of hosts and vectors through different population compartments after infection with a pathogen. It is assumed that the vector has no immunity to the pathogen and does not recover from infection. Births and natural deaths are not shown.

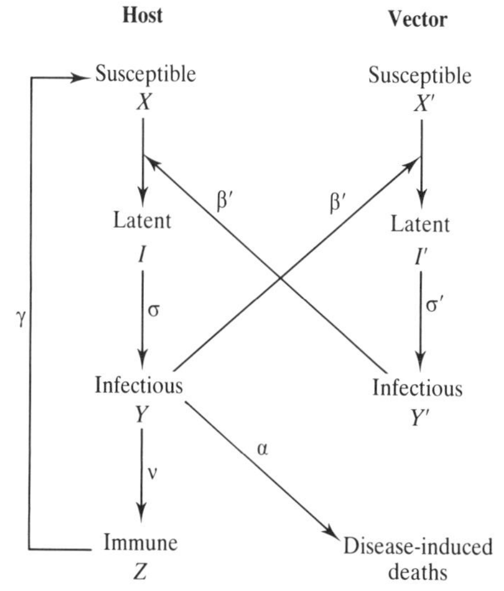

The equation shows that the number of secondary infections (R') depends on the ratio of vectors to hosts (N'/N) rather than the absolute abundance, or density, of either and is also related to the square of the transmission coefficient (β'^2). The above model can be extended or altered to reflect characteristics of particular host–vector–pathogen interactions, as noted by Anderson (1981, 1984) and to diseases with vectors and several host species (Rogers, 1988). For example, for rabbits and myxomatosis, the vector can be a mosquito and as the mosquito is a mechanical vector there is no latent period ($I'=0$), so dI'/dt and σ' disappear from the equations. Hence, the basic reproductive rate is then:

$$R' = \sigma\beta'^2(N'/N)/((\sigma+b)\ (\alpha+v+b)b') \qquad (5.64)$$

Brothers *et al.* (1982) discussed biological control of rabbits on subantarctic Macquarie Island, and considered that there was no estimate of how many fleas were needed to act as an effective vector of myxoma virus. The modelling suggests the higher the number of vectors the better, as that will increase the ratio of vectors to hosts (N'/N). However, the equation for the basic reproductive rate (equation 5.64) estimates the number of secondary infections, which is not necessarily the number of deaths of rabbits. Cooke (1983) reported results of an experiment in South Australia in which a vector of myxoma virus, a rabbit flea, was introduced into different rabbit populations. In one population non-infected fleas (X') were introduced and in another population infected fleas (Y') were introduced. An experimental control (no fleas, no myxoma) was also used. This is a rare example of an experimental study of this type. Survival of rabbits was lowest where infected vectors (Y') were introduced. The study could be extended by an experimental test of the relationship between infected vectors and mortality. Further details were described in section 3.11. Trout *et al.* (1992) reported results of a similar experiment involving flea release but also experimentally reduced flea abundance by use of an insecticide and vaccinated rabbits against myxoma virus. The study used one site and sequential experimental controls.

The model for a single host species can be extended to include multiple host species. Now transmission can occur between and within host species. Such situations may be more common than reflected in current models. For example, foot and mouth disease in Zimbabwe can occur in cattle and buffalo (*Syncercus caffer*) in the same area (Taylor & Martin, 1987). Sometimes the concern about alternative hosts requires further study. Seaman, Boulton & Carrigan (1986) reported that a viral disease, encephalomyocarditis, of domestic pigs was associated with a plague of rodents, house mouse and *Rattus* species, in a part of south-eastern Australia. Detailed mortality figures for pigs were reported but no evidence was presented that the rodents were or had been infected on or around the piggeries. The association could have been established or refuted by serological testing of the rodents, post-mortem examination for disease signs or pathogen isolation to determine their disease status.

The spatial spread of a disease in a wild host population can be modelled by extending the above equations to include movement, as noted in section 5.3.3. The application of spatial modelling to the control of a disease is described in section 6.8 as are other aspects of disease control.

The modelling and further study of diseases in vertebrate pests requires careful tests of models by comparing them with field data. Some suggestions for such tests are listed in Table 5.5. The tests are required in order to establish whether the structure of the models is appropriate, whether the estimates of parameters are biased and, finally, whether the predictions occur.

5.7 Rodent damage control

Statistical analysis of rodent damage was discussed in section 2.6 and control of such damage in section 3.5 and 3.17. Economic analysis was discussed in section 4.6.

There are many models of rodent population dynamics and many have been referred to earlier in this chapter. Hence, this brief section will not describe or review them again. Models of coypu (Gosling *et al.*, 1983) and mice (Singleton, 1989) dynamics were described in section 5.3.1 and those for cotton rats (Montague *et al.*, 1990) are referred to at the start of Chapter 6. Other general studies are by Stenseth (1977), Caughley & Krebs (1983) and Krebs (1988). There is scope for modelling of rodent damage to complement the models of rodent dynamics.

Table 5.5. *Tests of models of infectious diseases in vertebrate pests*

1. Assumptions in model structure
   - (i) Model component not included
   - (ii) Model component included when not needed
2. Parameter estimates
   - (i) Estimates biased because of data
   - (ii) Estimates biased because of estimation procedure
3. Predictions
   - (i) Estimate of threshold host density (or abundance)
   - (ii) Estimate of basic reproductive rate
   - (iii) Estimate of equilibrium host density (or abundance)
   - (iv) Estimate of equilibrium disease prevalence
   - (v) Estimate of trends in host abundance, abundance of infectives or deaths
   - (vi) Estimate of response to pest control

5.8 Bird strikes on aircraft

Statistical analysis of bird strikes was discussed in section 2.7 and control of such strikes was discussed in section 3.18. Economic analysis of strikes was very briefly discussed in section 4.7.

A computer model of three-dimensional flocks of 'prey' (birds) and their 'predators' (aircraft) is described by Major, Dill & Eaves (1986). The predators were modelled after the frontal area of aircraft engines and the prey were modelled after flocks of dunlin (*Calidris alpina*), a small wader. The model assumed the dunlin were randomly distributed in three-dimensional space and the aircraft of specified sizes and configurations moved near or through the flock. The predator speed and angle of elevation, and prey flock size, speed, length and height were systematically varied in the simulations to estimate the mean and variance of the number of bird strikes. The number of simulated prey 'captured' by the predator increased linearly with increasing number of prey (flock density). One limitation of the model, as noted by the authors, is the assumption of random distribution of birds, which is probably rarely true.

5.9 Bird damage to crops

Statistical analysis of bird damage to crops was discussed in section 2.8 and analysis of control in section 3.19. Economic analysis was discussed in section 4.8.

Wiens & Dyer (1975) developed a model of blackbird damage to corn by estimating the food requirements of bird populations. Food needs were estimated from equations of the energy requirements for birds of specified weights. The method requires estimates of the total bird population for each set of predictions. Weatherhead *et al.* (1982) described a method based on estimating bird populations and using the energetics approach of Wiens & Dyer (1975). Field sampling of damage is then not necessary. Damage estimates from an energetics model and from an enclosure study were similar (1.675 million kg and 1.681 million kg respectively). Both were substantially below a reported estimate of 113.586 million kg, though how this higher estimate was obtained was not described other than to note that there was no extensive fieldwork.

Woronecki *et al.* (1980) concluded that the energetics model of Wiens & Dyer (1975) underestimated blackbird damage. The damage occurred when the corn kernels were only 20–40% of their final weight. Otis (1989) reported that the energetics model had three particular problems. Firstly, the modelling required estimates of model parameters and these may not always be available. Secondly, the models were deterministic so that one input gave one output and this did not reflect the natural variation in damage observed. Thirdly, modelling is partly subjective so different modellers will produce different models and hence different predictions of crop damage. The comments of Otis

(1989) are valid and can be directed at any of the models described in this chapter and the next. Because of those concerns there is a need to compare predicted and observed results and test the differences.

5.10 Rabbit damage

Statistical analysis of rabbit damage was discussed in section 2.9 and control in section 3.20. Economic analysis was discussed in section 4.9.

Myers (1986) presented a graphical model of the eruptive dynamics of rabbits in Australia since the middle of the 1800s. The herbivore population (rabbits) increased, which decreased their food supply, and hence food for domestic livestock, and then both rabbits and food showed damped oscillations towards an equilibrium. Any balance was disrupted by the release of myxoma virus, which depressed rabbit abundance and increased food supply. In the graphical model rabbit abundance declined slowly after myxoma release. This contrasts with the sharp decline reported at the site of release (Myers *et al.*, 1954), though as myxomatosis probably did not spread to all parts of the rabbit's distribution, especially in the arid zone, the model may be realistic. Pech *et al.* (1992) presented a verbal and graphical model of predator (fox, cat) regulation of rabbits in semi-arid Australia. The model was an elaboration on an earlier graphical model of Newsome, Parer & Catling (1989), and suggested that when rabbits are not abundant their populations could be regulated by predators but when rabbits are abundant they are regulated by food not predators. The dynamics of rabbit populations were modelled by Dwyer *et al.* (1990) by assuming per capita birth and natural death rates to be independent of rabbit density. Seasonal changes in reproduction were incorporated into the model.

5.11 Erosion and vertebrate pests

Feral pigs, rabbits, feral horses, deer and other vertebrates are often accused of causing soil erosion. There are usually no field or experimental data to support the accusation. The relationship between ground disturbance, grazing and soil erosion can be modelled to assist with planning experimental studies to test associations.

Thornes (1988, 1990) modelled erosion as the rate of loss of soil, in terms of either the loss in kg/m^2/year or cm of topsoil/year. The rate of erosion ($\mathrm{d}E/\mathrm{d}t$) was assumed to be a function of the surface runoff per unit width (q) and slope (s).

$$\mathrm{d}E/\mathrm{d}t = k\, q^m s^n \qquad (5.65)$$

where k, m and n are constant model parameters to be estimated empirically. Surface runoff (q) was assumed to be determined by many factors, such as the depth of soil previously lost, the depth of remaining soil, an index of stoniness of the soil, vegetative cover (%), the runoff on bare soil, an effect of soil compaction and the rate of mixing of organic matter by physical and biotic processes. Vertebrate pests will presumably not influence slope (s) but could influence surface runoff (q) by influencing one or more of its determinants.

Herbivorous animals could contribute to erosion in several ways (Thornes, 1988, 1990). Firstly, grazing reduces vegetation biomass and hence vegetative cover. Secondly, herbivores can dig, scratch or root up the ground, which can have a direct effect on reducing vegetative cover. Erosion increases quickly when the vegetative cover is less than about 30% since cover (x) is related to erosion (y) by a negative exponential relationship (Fig. 5.16a) (Francis & Thornes, 1990).

Thornes (1988, 1990) assumed the rate of change of biomass of the ungrazed vegetation to be logistic, as assumed by Caughley (1980), Caughley & Lawton (1981) and Hone (1988b) for interactive plant–herbivore systems. This aspect of the modelling would need changing in environments where the rate of vegetative growth is determined more by rainfall and less by existing biomass. Robertson (1987) described such effects for a semi-arid area in Australia. The models of Thornes (1988, 1990) assumed that vegetative biomass was reduced by grazing. The biomass of vegetation was converted to percentage cover for modelling. Details of the models differ between Thornes (1988) and Thornes (1990). The former assumed a linear functional response by grazing herbivores and the latter study assumed a curvilinear response because of a saturation relationship between intake and availability of food. The conclusions were qualitatively similar, however.

The modelling of Thornes (1990) predicts that a given level of plant biomass corresponds to only one level of soil erosion, but a given level of soil loss, within limits, corresponds to two levels of plant biomass. Hence, erosion could be very low when biomass is high, because of high vegetative cover, or when biomass is low, because there is very little soil remaining that can erode. The predicted effect of grazing is to reduce the range of combinations of vegetative cover and soil loss which will change to a stable equilibrium of high cover and low erosion (Thornes, 1988).

Several studies in Great Smoky Mountains National Park in Tennessee provide data relevant to the topic of soil disturbance and erosion. Bratton (1974) reported that ground rooting by wild pigs in gray beech forest lowered the cover of understorey plants from 56–87% on undisturbed sites to 13–23% on disturbed sites. The reduction in cover and disturbance of the leaf litter were reportedly accompanied by soil erosion though the erosion was not estimated. Such low vegetative cover should result in erosion if the results of Francis & Thornes (1990) are general. Lacki & Lancia (1983) reported that ground rooting increased soil organic matter and acidity compared with non-rooted sites in gaps in beech forest in the park. The organic matter increased because of mixing of soil with surface litter. The effects of such mixing were included in the model of Thornes (1990) for prediction of the rate of erosion. The predicted effect of increased mixing was to decrease erosion (Fig. 5.16b). The slope of the gap sites ranged from 17% to 60% (Lacki & Lancia, 1983) so erosion would be expected to occur.

Singer, Swank & Clebsch (1984) showed that in areas of intensive pig rooting the bulk density of soil was significantly lower, the percentage of bare ground was significantly higher but sediment yield was not significantly different to that in non-rooted areas, though it was slightly lower in the rooted area. Pig rooting apparently increased water infiltration so sediment yield did not increase. The comparison of bare ground used a sequential test of the same sites six years prior to and eight years after occupation by wild pigs. The other comparisons used simultaneous tests of different areas. Each test of difference was a Student's t test.

Caughley (1970b) and Caughley & Lawton (1981) suggested that non-territorial ungulates exhibit an eruptive fluctuation in population size over time since introduction. The trend resulted from the effect of ungulates on their food supply and was derived from a two-trophic model of population dynamics for plant and herbivore populations, similar to those described in section 5.3.1. Feral pigs are non-territorial ungulates, which can reduce the abundance of their food supply (Howe, Singer & Ackerman, 1981). Challies (1975) reported that feral pigs on Auckland Island, New Zealand, appeared to have an eruptive fluctuation over time since introduction. Singer (1981) suggested that feral pigs in Great Smoky Mountains National Park, Tennessee, erupted in abundance and then stabilised. In contrast, Crawley (1983) considered that the effects of

Fig. 5.16. Predicted relationships between the rate of soil erosion and (*a*) vegetative cover and (*b*) mixing of soil by foraging animals. (After Thornes, 1988, 1990.)

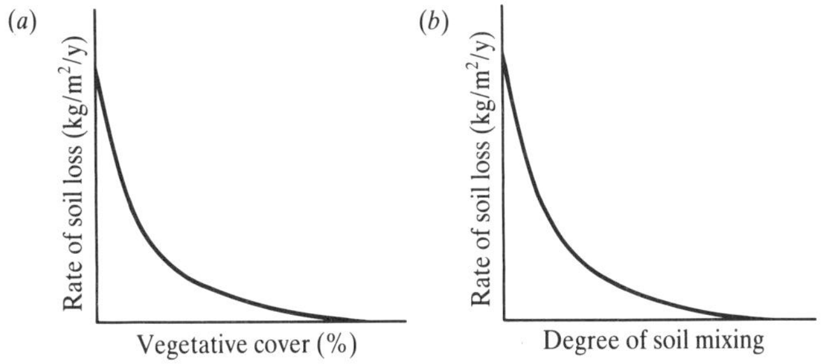

rooting by feral pigs decreased the abundance of vegetation but had negligible effect on the pig population. If ground rooting decreases the food supply to pigs then the extent of new rooting by feral pigs is negatively related to food availability, and the extent of new rooting is predicted to be related to pig abundance in a curvilinear manner (Hone, 1988b). The extent of erosion would also be predicted to be related to abundance in a curvilinear manner.

The predicted relationship between the extent of new rooting and pig abundance suggests that the assumption of Belden & Pelton (1975), Conley (1977) and Giles (1980) of a simple positive relationship between pig abundance and rooting extent may have been simplistic, as some levels of rooting correspond to more than one level of pig abundance. The model also indicates how the disparate results in Hawaii of Cooray & Mueller-Dombois (1981) and Ralph & Maxwell (1984) differed in the relationship between rooting and pig abundance. Each study may have been measuring rooting and pig abundance at different times since introduction of feral pigs or each population may have experienced different control regimes leading to different pig densities. The studies measured extant rooting rather than the extent of new rooting, so the conclusions of the modelling are tentative.

The broader topic of vertebrate pests and soil erosion needs further good experimental studies to test the predictions of the modelling. Currently there are insufficient data to estimate parameters of the erosion models.

5.12 Conclusion

The modelling of population dynamics and damage is a mix of theory, mathematics and field data. The modelling is an encouraging start to the problem of describing the dynamics of change and patterns in vertebrate pest abundance or damage. There is a great need for testing of the assumptions and predictions of the models, and hence a closer link between the theory and field data. Agreement between observations and predictions is encouraging, however, it should be considered carefully. The agreement does not prove the model is correct. In philosophy this mode of reasoning is called the fallacy of affirming the consequent (Hempel, 1966). Mollison (1991) argued that obtaining good agreement of predictions with data may not always be a sound basis for extrapolations and predictions. Falsification may be more informative. That will certainly be true of tests of models but is not of much short-term use to farmers or wildlife managers who have problems with vertebrate pests. In the longer-term, sound testing of models will satisfy both modellers and managers.

There is also a need for closer links between the modelling described here and the economic analyses in Chapter 4. Table 5.6 is a start at getting closer links by listing relevant questions. Walters (1986) discusses similar ideas on judging model credibility and is recommended reading. He describes repeatability, robustness and completeness of models. Modellers have a choice between developing a specific model for each pest population or disease, or using a set of generic models that can be adapted quickly to a particular pest or disease. Shrum & Schein (1983) suggested that generic models were more flexible than specific models in planning responses against plant pests exotic to the U.S.A.

The volume of modelling literature on the topics in this chapter is not proportional to the volume of the statistical or economic literature on the same topics. In particular, there is great disparity for the topic of livestock predation: a large literature on damage and a modest literature on economics and seemingly nothing on modelling. In contrast, there is a large literature on modelling infectious diseases but very little on economic analysis and a small literature on statistical analysis.

The modelling of control of damage and vertebrate pests is a similar mix of theory and data and is the topic of the next chapter.

Table 5.6. *Checklist of questions to be answered in formulating and testing models of population dynamics and impacts of vertebrate pests*

1. Does the model structure reflect the field situation?
2. What are the consequences of departures in model structure from the field situation?
3. What are the assumptions in the model?
4. Have the assumptions been tested or are they based on laboratory or field data?
5. What are the consequences of small or large departures from the assumptions?
6. Do the methods of parameter estimation give unbiased and precise estimates?
7. What are the consequences of small or large biases in parameter estimates in terms of accuracy, precision, repeatability and robustness of predictions?
8. What are the qualitative and quantitative predictions of a model?
9. How can the predictions be tested?
   - experimental or inferential studies
   - statistical test for goodness of fit
10. Given the results of past tests of a model, how can the model be altered to improve the accuracy of its predictions?
11. Does the model include any economic principles or analysis?
12. Can the field data be more accurately or precisely predicted by an alternative analysis?

6

Modelling of control

'What could modelling possibly tell me about control? All I have to do is go in and blow them away.' Such a view is concise and clear but short-sighted. Modelling can assist with planning of control. For example, a mathematical model of the population dynamics of the cotton rat (*Sigmodon hispidus*) in sugarcane crops in Florida was used to evaluate the effects of different timings of rodenticide applications on cotton rat abundance (Montague *et al*., 1990). It was concluded that a single application at least eight months after harvest or double applications at eight and ten months after harvest would be more effective in reducing future damage than earlier rodenticide treatment. The modelling suggests a control action that can be tested empirically.

This chapter is about modelling the processes of control in vertebrate pest control. These models describe patterns and processes in the methods, rather than just the effects of the methods. The models are often based on analogy with ecological processes such as predator–prey and plant–herbivore relationships. The results can be of strategic use in planning control.

The literature emphasises modelling the effects of pest control rather than modelling the effects of damage control. That will be reflected in section 6.1 in the theory and case studies examined; however, I suggest it is not the most efficient or economic approach in vertebrate pest control. There is a need for more modelling of damage control. Spatial and temporal aspects of modelling are discussed in section 6.2 and in sections 6.3 to 6.13 the processes of control rather than simply the effects on pest abundance are modelled.

Krebs (1985) described four principles of exploitation of wild populations that apply equally to harvesting and control. Firstly, exploitation reduces abundance of a population. Secondly, populations compensate for exploitation if the exploitation occurs at less than a certain level. Thirdly, exploitation can push a population to extinction. Fourthly, somewhere between exploitation and extinction is a level of the maximum sustained exploitation or yield. Some of the modelling of control in this chapter is concerned with modelling or estimating that level of exploitation.

6.1 Models of the response of pest populations to control

There is a sizeable literature on modelling hunting of game populations and populations from which animals are or have been removed. The examples range from ducks to whales. Caughley (1980) has stressed the similarity of the removal processes in hunting and pest control. The similarities will be examined further here.

Models of the effect of control on pest abundance usually have a general structure of the type:

$$N_t = f(N_0) - P\,g(C) \qquad (6.1)$$

where N_t is pest abundance at time t, $f(N_0)$ is a function for pest abundance at time 0 in the absence of pest control, P is the number of people and $g(C)$ is a function for pest control per person and is equivalent to the functional response. Equation 6.1 is simply a model for population growth like equation 5.16, modified to include losses from the population because of pest control. The equation says nothing about the type of control other than it is lethal.

Any method of control aiming to reduce pest abundance must be used at a frequency and intensity that depresses the population rate of increase. The rate of offtake of pests must exceed the intrinsic rate of increase, for abundance to decline (Singer, 1981). That is:

$$N_t < f(N_0) - P\,g(C) \qquad (6.2)$$

When animals are removed from a population, such as during hunting or control, then the net growth may be described by the exponential equation. Eberhardt (1987) reported encouraging results of fitting an exponential function to temporal changes in abundance of 13 large mammal species. The exponential rate of increase was estimated using three analyses; log-linear least squares regression, non-linear least squares regression and the ratio method. Each method gave very similar estimates of rate of increase. Exponential growth occurred when substantial numbers of animals were removed from each population. If the known number of removals (C) occurs just before an annual census then population growth can be described by:

$$N_t = N_0\lambda^t - C \qquad (6.3)$$

If the known removals occur just after the previous census then:

$$N_t = (N_0 - C)\lambda^t \qquad (6.4)$$

where N_t is the number of animals at time t, N_0 is the number at time zero, λ is the finite rate of increase and C is the number of animals removed. Eberhardt (1988) described the timing of removals differently, as occurring just after annual reproduction (equation 6.3) and just before annual reproduction (equation 6.4). It is assumed that the C animals are removed in a short period of time. The number of removals can be known accurately when trapping and shooting occur but not usually when poisoning occurs. Eberhardt & Pitcher (1992) applied equation 6.4 to analysis of caribou (*Rangifer tarandus*) data from Alaska and equations 6.3 and 6.4 to wolf (*Canis lupus*) data. Crawley (1983) gave similar equations though applied to logistic growth with removals.

The general function for control ($g(C)$) in equation 6.1 may take many forms. Table 6.1 lists some control functions with comments on the form or type of control activity. Many of the control functions come from fisheries and game management. The equations in Table 6.1 are illustrated in Fig. 6.1. The predicted relationship between numbers killed and effort could be also graphed and would show a similar variety of shapes. Many other forms of the control

Table 6.1. *Forms of the general control function* (g(C)) *that could be used in equation 6.1. In the equation* n *is the number of pests killed,* E *is the effort,* N *is the number of pests before control and* a, b *and* c *are constants to be estimated empirically*

Number	Equation ($g(C)$)	Comment
1.	n	Constant removals (constant quota)
2.	aE	Constant effort (equals model 8 when $c=$o)
3.	$N-n$	Constant remainder
4.	aN	Constant fraction removed
5.	aEN	Surplus yield
6.	aE^bN	Surplus yield with saturation on effort when $0<b<1$
7.	$aN-b$	Linear regression
8.	aE^bN^c	Surplus yield with parameter c not equal to one (Cobb–Douglas function)
9.	$a(1-e^{-bN})$	Saturation on number
10.	$a(1-ce^{-bN})$	Saturation on number, with extra parameter c
11.	aN^{-b}	Constant product yield
12.	$N(1-e^{-aE})$	Catch equation based on random search
13.	aEN (when $N<N_t$) n (when $N>N_t$)	Ramp model where N_t is a threshold
14.	$n(t)$	Variable with mean n, variance s^2

function could be used, such as the various equations of the functional response listed in May (1981b), Eberhardt (1988) and Lundberg & Danell (1990). They are mostly for use in differential equations describing the rate of change of populations.

The exact form of $g(C)$ used in a model will depend on the assumptions or the administrative framework that applies to the control. A detailed comparison of the biases and robustness of the analyses has not been reported. Constant removals (expression 1 in Table 6.1) occur when a constant number of pests are killed per year. Pest control can become a constant harvest when a quota is applied to the number of pests killed. Constant effort (expression 2) describes pest control where the inputs are similar between years or locations. This often happens because of limits on money and staff. A third strategy (expression 3) is to leave a constant number of pest animals after control. This is the constant remainder policy and is often done to attempt to ensure a continued harvest. The constant harvest policy is the most destabilising for the harvested population as harvest continues even if abundance declines, such as in a drought or cold winter. A constant effort policy is less destabilising and a constant remainder policy less destabilising still (Clark, 1981). Anderson *et al.* (1981) described a framework for assessing the effects of constant effort and

Fig. 6.1. The predicted relationship between the number of pests killed by control, and pest abundance before control for various forms of the control function ($g(C)$). The numbers 1–14 refer to equations listed in Table 6.1.

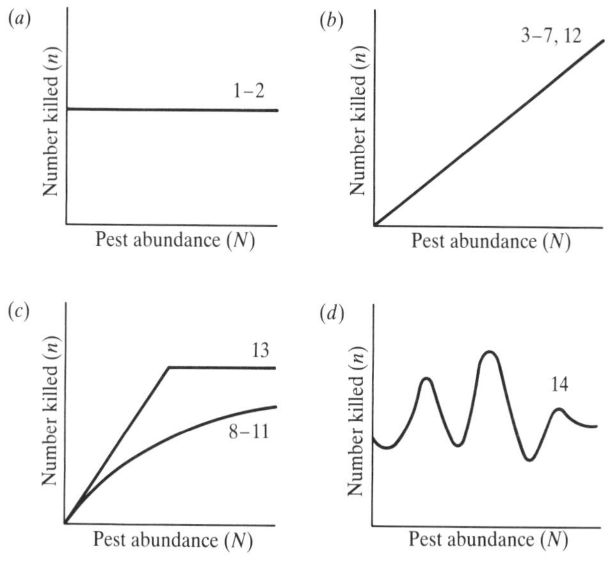

constant harvest policies for the control of rabies in foxes though did not describe the results of such an analysis.

Clark (1981), Krebs (1985) and Begon *et al.* (1990) described the use in fisheries management of the surplus yield model (expression 5). In expression 5, a is the proportionality coefficient or catchability coefficient. The ratio n/E is defined as the catch per unit effort (CPUE) and has been used as an index of wildlife abundance (Caughley, 1980). In vertebrate pest control this relationship is used, though invariably implicitly – without knowing. A piece of modelling trivia is that the surplus yield model (aEN) is based on the same assumption of mass action as that described in section 5.6 for changes in the number of hosts susceptible to an infectious disease and the linear functional response in predator–prey models. Caughley, Dublin & Parker (1990) and Basson, Beddington & May (1991) described a harvest model, EN, which is equal to the surplus yield model when $a=1$. Their models were used to describe the rate of harvest of ivory.

Boulton & Freeland (1991) used expression 7 to analyse data for shooting of buffalo in northern Australia. The measure of the number of buffalo shot was the daily tally. Expression 8 was described by Walters (1986) as a general production function in which the parameters b and c account for the effects of non-random search and saturation of harvesting gear. The constant product yield model (expression 11) was used by Basson *et al.* (1991) to describe harvest of ivory. Fryxell *et al.* (1991) used the catch expression (12) in a study of hunting of white-tailed deer (*Odocoileus virginianus*) in Ontario. Expression 14 is probably very common but is almost never described. The frequency distribution should be described, as well as the mean and variance. Gosling & Baker (1989) reported results of variable effort in trapping coypu in south-eastern England. They described a significant multiple linear regression between the annual change in coypu abundance (y) and trapping intensity (x_1) and the severity of winter (x_2).

The two parts of equation 6.1 can be graphed (Fig. 6.2) to estimate the population size that will occur for a given level of hunting (N_p). In Fig. 6.2 it is assumed that the pest population shows logistic growth and that there is a linear relationship between population size and offtake (expression 5). The relationships shown in Fig. 6.2a can be expressed mathematically. The logistic growth rate of the pest population can equal the rate of harvesting at equilibrium:

$$rN(1-(N/K))=aEN \tag{6.5}$$

where K is the carrying capacity and r is the exponential rate of increase. The population size at this equilibrium (N_p) can be estimated by rearranging the equation. The term, N, cancels out from both sides of the equation, and dividing both sides by r gives:

$$1-(N/K)=aE/r$$

This can be rearranged to give:

$$N=K-(aEK/r)=N_p \tag{6.6}$$

Equation 6.6 shows that the equilibrium population size will increase if K or r increases but size will decrease if a or E decreases. The population can be driven to extinction ($N_p=0$) when:

$$N_p=K-(aEK/r)=0$$

That is, after rearranging and cancelling K, when:

$$E=r/a \tag{6.7}$$

If hunting takes a constant number, n, of pests and population growth is logistic (Fig. 6.2b) then:

$$rN(1-(N/K))=n \tag{6.8}$$

This equation has no single solution for N, except when $N=K/2$. Otherwise there are two values of N which correspond to the equilibrium population size.

Control by hunting can be described as by Smith (1985). Assuming logistic

Fig. 6.2. The relationship between a logistic model of pest population growth (solid line) and the (*a*) surplus yield model of the rate of hunting (dashed line) and (*b*) constant number offtake (dashed line). The point(s) of intersection determines the pest population size (N_p).

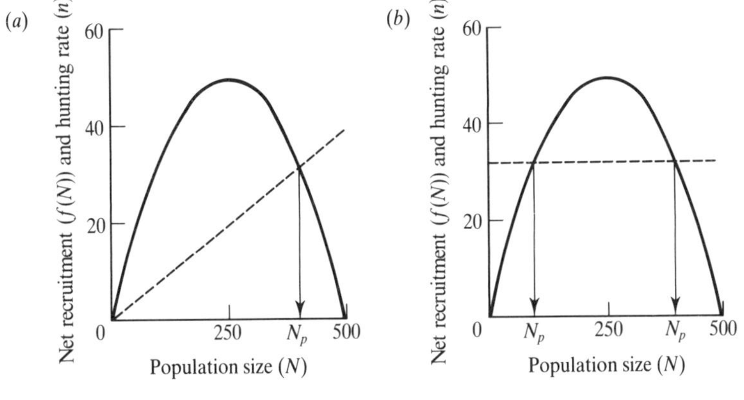

growth of the pest population, and a constant per capita mortality rate, a, from culling (expression 4) then:

$$dN/dt = rN(1-(N/K)) - aN \qquad (6.9)$$

At equilibrium $dN/dt=0$, so after rearranging:

$$N = K - (aK/r) = N_p \qquad (6.10)$$

This equals equation 6.6 when $E=1$.

As hunting effort increases the hunting line in Figure 6.2a becomes steeper, so the pest population size decreases. At a much higher hunting rate the only pest population size at which the growth rate is equal with the hunting rate is a population of zero. The population has been hunted to extinction. The same effect can be achieved by a technological change that increases the efficiency (a) of a control method. This is simplistic, of course, as the hunting efficiency would change as the population became scarce. The hunting equation may not be linear, as discussed by Clark (1981) and expressions 8 to 11 suggest. In theory, the relationship could also be disjointed so it is linear and increasing at low pest abundance, but constant (horizontal) at high pest abundance. This is the ramp model (expression 13).

If an analogy is drawn between a parasite in an environment of a host and a pest in the environment then another analysis is suggested. The hunting rate of pests, a, can be expressed differently based on the approach described by Anderson (1982d) for control of hookworms and roundworms. Obviously, they are not vertebrate pests but the principles are the same. The hunting rate, the death rate of pests, is equal to:

$$a = -\ln(1-gh) \qquad (6.11)$$

where g is the proportion of a population treated by control and h is the proportional kill of the treated population. The equation assumes that areas to be treated are selected at random. By rearrangement:

$$h = (1-e^{-a})/g \qquad (6.12)$$

The relationship predicts an increase in the proportional kill of the treated population as the hunting rate increases but the increase is at a decreasing rate (Fig. 6.3).

Anderson (1982d) suggested that a more efficient control method is to select areas of highest pest abundance and concentrate control there. This analogy highlights the idea of the frequency distribution of parasites per host, which is analogous to the frequency distribution of pests or damage per area. The latter were described in Figs. 2.1 and 3.1, as well as the response to control (Fig. 3.1). The frequency distribution of parasites per host is usually highly skewed (Anderson, 1982d) and if the frequency distribution of damage or pests is similarly highly skewed, as described in section 2.2, then most control, if applied at random, will produce little change in damage or pest abundance. The herbivore damage model of Crawley (1983), which incorporated information on the frequency distribution of damage, could be used to model the response of damage to spatial variation in control.

Modelling has been used to estimate the response of populations to control or eradication efforts. Caughley (1980) estimated the effect of control on the abundance of a hypothetical pest population. It was assumed that the population, in the absence of control, increased in abundance over time according to logistic growth, carrying capacity (K) was 1000 animals and the intrinsic rate of increase (r_m) was 0.6/year. When a constant number of animals were removed annually (expression 1) the population stabilised at a new density, unless the number removed exceeded the maximum sustained yield (150 animals/year $= r_m K/4 = 0.6 \times 1000/4$). If 200 animals were removed annually the operation took over 10 years to completely eradicate the pest.

When a constant proportion of the population was removed annually (expression 4) the population also stabilised at a new equilibrium, unless the rate of removal exceeded the maximum rate of 0.3. At a constant removal rate

Fig. 6.3. The predicted relationship between the proportional kill of a treated population and the hunting rate, when hunting occurs at random with respect to pest abundance as described in equation 6.12. The effect of varying the proportion of a population treated by control (g) is also shown.

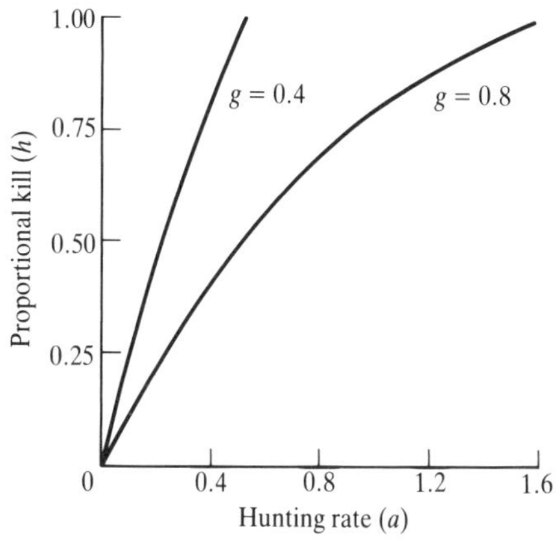

of 0.7 the eradication took 42 years. Caughley (1980) noted two limitations to the modelling. Firstly, it was assumed that the reduction in population size did not result in an increase in the rate of increase of the population triggered by the extra resources for survivors. Secondly, the models assumed that a unit of hunting or control effort removed a constant proportion of the population. Animals are unlikely to be equally catchable, especially when abundance is very low and the population is being hunted.

Caughley (1980) described a model of hunting of herbivores by humans. The model had three trophic levels. The herbivores had dynamics determined by their food supply. In contrast, the humans had no dynamics. The model predicted a lower level of sustained harvest from the herbivore population than if it had logistic growth. Hence, the population with a variable food supply will be easier to control.

Stenseth (1977) modelled the extinction of populations rather than the dynamics of individuals within populations. If there are X populations then the rate of change of populations was assumed to be:

$$\mathrm{d}X/\mathrm{d}t = mX(1-(X/T)) \tag{6.13}$$

where m is the effective dispersal rate and T is the maximum number of populations in a habitat. This is a form of logistic growth. If populations are becoming extinct at a rate cX then:

$$\mathrm{d}X/\mathrm{d}t = mX(1-(X/T)) - cX \tag{6.14}$$

The equilibrium number of populations (X_e) is given by setting $\mathrm{d}X/\mathrm{d}t = 0$, and rearranging to give:

$$X_e = T(1-(c/m)) \tag{6.15}$$

Hence, the equilibrium number of populations increases with decreasing extinction rate (c) and increasing dispersal rate (m). If a control objective is to decrease the equilibrium number of populations (low X_e) then the extinction rate (c) must be increased, the maximum number of populations (T) must be reduced or the dispersal rate (m) must be decreased.

Eltringham (1984) described six methods of estimating the maximum sustainable yield for a variety of wildlife populations. If that yield is the threshold below which a population will be pushed to lower levels, and eventually to extinction, then the methods could be used to estimate a harvest rate which can then be used as a control rate. Caughley (1980) showed, by modelling, the effect of the differential harvesting of each sex in a wildlife population. As the proportion of males in the harvest increased, above 0.5, then the sustainable yield by harvest increased. The implication here is that control should reduce the proportion of males removed from a pest population when the opportunity for such action occurs. The surviving pest population will then have a low proportion of females, the effect of which will be to lower the sustainable yield, that is, make the population more vulnerable to the effects of control.

Links between breeding and control can be examined. Continual control of a pest population may alter the numerical response (Fig. 6.4). Control, if continued, could decrease the intrinsic rate of increase (r_m) of the population. Control may also shift the numerical response relationship to the right so that for the same level of food availability the population has a lower rate of increase.

6.2 Spatial and temporal aspects of control

6.2.1 *Spatial*

If immigration is a significant component of the increase of a pest population, then that movement should be recognised in planning control. Stenseth (1981) and Stenseth & Hansson (1981) analysed, by modelling, the role of immigration in population dynamics and pest control. They concluded that the best control strategies vary with the dynamics of the population. If empty patches (no pests) occurred and methods were available to prevent

Fig. 6.4. Possible effects of continual control on the numerical response of a pest population to changes in its food supply. The solid line shows the response before control and the dashed line shows the response after control.

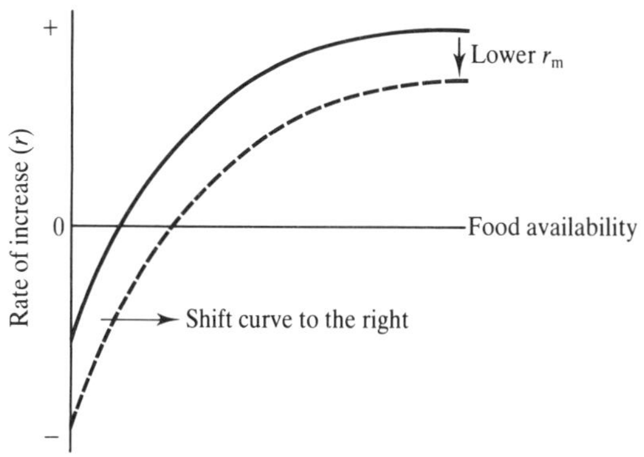

immigration, then those methods should be used rather than try to increase mortality elsewhere in the population. This is the strategy used to control damage by feral pigs in Hawaii Volcanoes National Park, Hawaii (Stone, 1985; Hone & Stone, 1989).

If effective methods for reducing immigration were not available and populations had low extinction rates, then effort should concentrate on influencing mortality and reproduction (Stenseth, 1981; Stenseth & Hansson, 1981). For populations with high extinction rates, efforts should concentrate on reducing immigration, though how this was to be done when no effective methods existed was not explained.

Buechner (1987) examined further, by modelling, the control implications of animal movements. Dispersal and density in habitat patches can be influenced by; (i) perimeter/area ratios, (ii) edge permeability, (iii) appearance and stopping rates of dispersers, (iv) habitat preferences of dispersers, and (v) the size of the population sink or pool. Changes in any of these factors have most effect on dispersal when the value of a factor is low compared with the range of values that it can have. If death rates in the source population were low, small increases in mortality, for example by trapping, may have a large effect on immigration. If mortality rates were moderate, increasing mortality further was predicted to have little effect on immigration. In this situation altering edge permeability, for example by fencing, may be more effective. Stamps, Buechner & Krishnan (1987) described the results of similar studies and concluded that when habitats have low edge permeability that is more important as an influence on dispersal than the perimeter/area ratio. Conversely, when habitats have higher edge permeabilities then emigration is determined more by the perimeter/area ratio.

6.2.2 *Temporal*

Vertebrate pest control is often implemented after damage is obvious and considered serious. That is not necessarily the most efficient or economic timing of control. Various situations can be distinguished which generate different timings of control. Firstly, if damage is expected then prophylactic control can be implemented. The decision to control pests is taken before damage is occurring or perceived. Secondly, if damage is uncertain, then the decision to control pests may be delayed. When the decision occurs will depend on the risk aversion of the decision maker (Norton, 1976).

The timing of damage by pests can indicate a solution to a problem. For example, if predation by a vertebrate pest is restricted to new-born lambs then predation losses could be reduced by timing lambing over a short time period so that the predators are swamped. Such a relationship has been suggested for natural predator–prey systems (Taylor, 1984). The response will only be obtained if there is no surplus killing by the predators.

In planning control in conservation areas, the frequency of vertebrate pest damage may be important. Frequent disturbance of ecosystems is considered to have a deleterious effect on species. However, models by Connell (1978) and Huston (1979) suggest that species richness or diversity of ecosystems may be directly related to the frequency of disturbance with richness or diversity highest at intermediate levels of disturbance. The role of pest species in creating such disturbances has not been reported.

Southwood (1988) reviewed these models and the broader role of habitat disturbance in influencing biological diversity and life-history strategies of species. Disturbance, varying from volcanic eruptions to the fall of a tree, acted as a templet for various life-history strategies. The relevance of this approach to vertebrate pest control may depend on the ability of a community of species to adapt to disturbance. The dramatic effects of introduced herbivores and predators on island communities (Pimm, 1987) may be a consequence of the limited time period the original communities and species have had to adapt to the new species. The capacity of such communities to return after the disturbance to their prior state is a measure of their stability (Pimm, 1984). When the impact of vertebrate pests can be regarded as a form of disturbance, then analysis of conrol may be assisted by using a framework to describe a disturbance regime. This regime concept was described by Karr & Freemark (1985) as being in four parts: the type of disturbance, the spatial and temporal aspects of the disturbance, the type of biological system in which the disturbance occurred, and the regional context relating landscape features to disturbance.

6.2.3 *Combined spatial and temporal aspects of control*

Spatial and temporal aspects of resource management have been combined in geographic information systems. An example is for fire management in Australia (Kessell *et al.*, 1984). The computer system uses primary site data such as elevation and aspect to predict site vegetation, and time since fire and vegetation type to predict fuel levels. Fire behaviour, such as rate of spread, can be predicted from weather and fuel data. Such systems offer considerable potential for improvement in the planning and evaluation of vertebrate pest control. The systems could use site data to predict damage or pest abundance, and models of population growth to describe temporal changes, both before and after control.

6.3 Poisoning

Statistical analysis of poisoning was discussed in section 3.5.

6.3.1 *Dynamics of poisoning*

The process of poisoning vertebrate pests can occur in two steps. Initially, pests are offered non-poisoned bait. This step is called free-feeding or pre-baiting. When removal of bait has reached a high and stable level, then the second step is instigated which involves switching poisoned for non-poisoned bait. A model of the two-step process was described by Hone (1992a) with the poisoning step emphasised in the outline here.

During free-feeding pests may be isolated (I) by geographical (rivers, mountains) or behavioural (neophobia) factors if they cannot find or eat the non-poisoned bait, susceptible (S) if they would eat the bait when they find it, or they eat the non-poisoned bait (E) (Fig. 6.5). When poisoned bait is offered some pests may be isolated (I) by geographical or behavioural factors (Fig. 6.5). A more detailed model would treat these factors separately and allow, where necessary, movement of pests between them. Isolated animals can become susceptible (S) at a per capita rate a. Susceptible animals can become poisoned (P) and after a latent period ($1/h$), during which poison signs are not apparent, they show signs and are then in the fourth and final compartment (Y) (Fig. 6.5). Animals which show signs die at a per capita rate j or recover at a per capita rate k and again become susceptible or recover at a per capita rate m and become isolated. During poisoning animals in each compartment can die naturally at a per capita rate d or births can occur at a per capita rate b. Total population size, or density, is N. The amount of poison in the environment (W) is determined by the rates of additions (A), weathering (μ) and removal by pests, at a per capita rate c.

In a model of infectious diseases, as described in section 5.6, the rate of change of the infected population is assumed to be proportional to the product of the number of susceptible animals and the number of infectious animals (Bailey, 1975). The obvious difference between the spread of infection or poison through a population is that in the simplest case, poisoning is analogous to a non-infectious disease. Individuals contract the disease (poison) by contact with poisoned bait, not by contact with infectious (poisoned) individuals. Hence, some of the basis assumptions of disease models need to be changed. The change in the number of pests eating non-poisoned bait (E) is proportional to the product of the number of susceptibles (S) and the weight of free-feed bait (B). The change in the number of poisoned animals (P) is proportional to the product of the number of susceptibles (S) and the amount of poison in the environment (W). The proportionality constant is c. Similar models of the dynamics of infectious diseases, wherein disease spread is determined by environmental sources of infection, were described by Anderson (1981), Anderson & Trewhella (1985) and Hochberg (1989).

The flow of individuals between compartments in a pest population (Fig. 6.5) can be described by a series of equations. In the equations births (b) and natural deaths (d) are not shown.

Free-feed period:

$$\mathrm{d}I/\mathrm{d}t = -\mathrm{a}\mathrm{I} \qquad (6.16)$$

$$\mathrm{d}S/\mathrm{d}t = aI - cSB \qquad (6.17)$$

$$\mathrm{d}E/\mathrm{d}t = cSB \qquad (6.18)$$

$$\mathrm{d}B/\mathrm{d}t = F - (cE + \mu) \qquad (6.19)$$

where F is the rate of adding free-feed and E is the number of animals eating the free-feed.

Poisoning period:

$$\mathrm{d}I/\mathrm{d}t = m\ Y - a\ I \qquad (6.20)$$

Fig. 6.5. Compartment model of poisoning a pest population. Arrows indicate direction of transfer and the associated letters indicate rates of transfer per unit time. Compartments show actual numbers of animals, and are those animals isolated (I), susceptible (S), eating non-poisoned bait (E), eating poisoned bait (P) and showing signs of poisoning (Y). Births and deaths are not shown. (After Hone, 1992a).

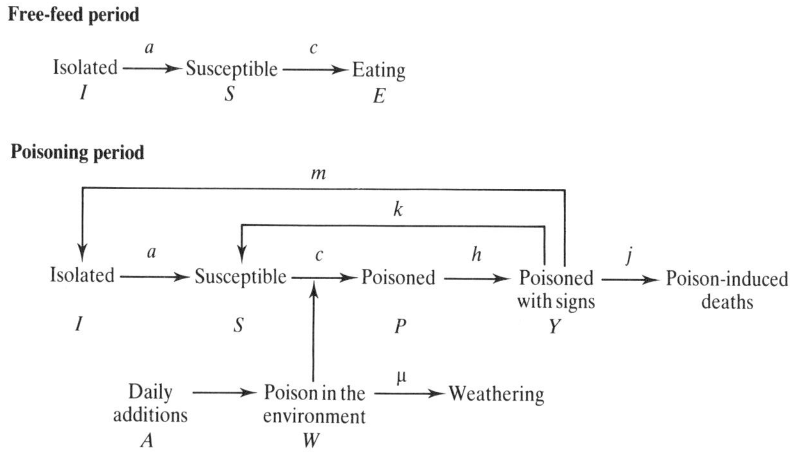

$$dS/dt = a\ I + k\ Y - c\ S\ W \quad (6.21)$$

$$dP/dt = c\ S\ W - h\ P \quad (6.22)$$

$$dY/dt = h\ P - j\ Y - k\ Y - m\ Y \quad (6.23)$$

$$dW/dt = A - (cN + \mu)W \quad (6.24)$$

The equations can be solved for the equilibrium situation wherein $dS/dt = dP/dt = dY/dt = 0$. There is no equilibrium abundance of pests above zero but there is a threshold amount of poison that must be added per unit time to achieve control or eradication. The threshold (A_c) occurs if an aim is to lower the number of susceptibles (S) to less than the number of pests poisoned (Y), so $S/Y < 1$. The threshold is:

$$A_c = (cN + \mu)\ (j + k + m)\ /\ c \quad (6.25)$$

The equations can be rearranged to estimate the rate at which pests die from poisoning (jY):

$$jY = ((cSA)/(cN + \mu)) - (k + m)\ Y \quad (6.26)$$

Equation 6.26 simply says that to maximise the rate of poison-induced deaths, the number of pests that pass from susceptible to poisoned with signs ($cSA/(cN + \mu)$) must be a maximum to the number of pests poisoned with signs that recover ($(k + m)Y$). The equation thus shows strategic options for control (Table 6.2).

The model predicts that pest abundance will decline after the start of poisoning, and the decline will be reversed sigmoidal – slow at first then more rapid then slower (Fig. 6.6). That corresponds to a sharp increase then decrease

Table 6.2. *Strategic options for increasing the effects of poisoning on a pest population by increasing the rate of deaths* (jY) *as described in equation 6.26. The options assume the number of susceptibles is held constant*

Number	Option
1.	Increase the rate of change of susceptibles to poisoned (c)
2.	Increase the rate of addition of poison (A)
3.	Decrease the rate of weathering of poison bait (μ)
4.	Decrease the rate of recovery of poisoned to susceptible (k)
5.	Decrease the rate of recovery of poisoned to isolated (m)

in deaths over time. Halpin (1975) described such a trend for deaths for poisoning of animals though did not model the process. Hone (1992a) described evidence to support the prediction for poisoning of rabbits. Taylor & Thomas (1993) presented data on trends in bait take by rats (*Rattus norvegicus*) during eradication from an island near New Zealand. The trend was reversed sigmoidal.

Pest animals such as mammals and birds have a maximum rate at which they can eat food, such as poisoned bait. A modification to the transmission term was described by Hone (1992a) as:

$$cSW/(1+gW) \tag{6.27}$$

where g is a saturation coefficient such that when $g=0$, then the transmission term reduces to that in the basic model.

Equations 6.21, 6.22 and 6.24 for the saturation model are now:

$$\mathrm{d}S/\mathrm{d}t = aI + kY - (cSW/(1+gW)) \tag{6.28}$$

$$\mathrm{d}P/\mathrm{d}t = (cSW/(1+gW)) - hP \tag{6.29}$$

$$\mathrm{d}W/\mathrm{d}t = A - (cN/((1+gW)+\mu))W \tag{6.30}$$

At equilibrium the threshold rate of addition of poison is:

$$A_c = (cN + \mu(1+gW))\ (j+k+m)/c \tag{6.31}$$

Fig. 6.6. Changes in the percentage of feral pigs that were alive since the start of poisoning with warfarin, as predicted by equations 6.33 to 6.36 (--------). The observed trend in the percentage of feral pigs alive is also shown (————). (After Hone, 1992a).

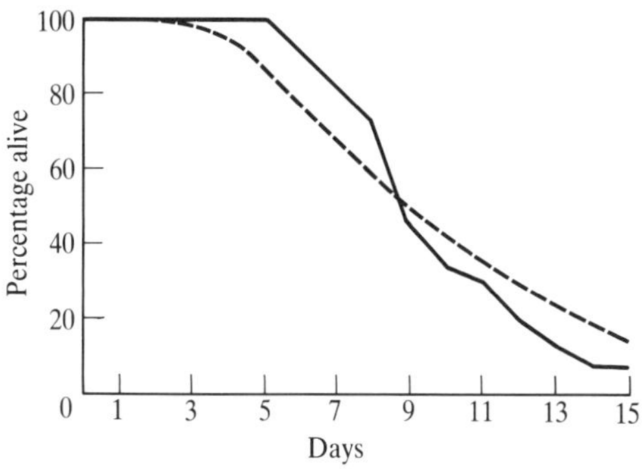

and the rate of deaths is:

$$jY = (cSA/(cN + \mu(1 + gW))) - (k + m)Y \qquad (6.32)$$

The effect of feeding saturation is to decrease the rate of deaths. When feeding saturation is absent ($g=0$), then equation 6.32 reduces to equation 6.26 and equation 6.31 reduces to equation 6.25. The basic model can be applied in different field situations and details were given by Hone (1992a).

The results of this modelling can be linked to modelling of chemical control of a pest population that is infected with an infectious pathogen or parasite. Carpenter (1981) modelled chemical control as simply being an extra additive source of mortality for the pest. The model predicted that pesticides permitted a less-virulent disease to control the pests. This prediction suggests that the observed reduction in virulence of myxomatosis virus in Australia (Fenner, 1983) may not have been solely a result of virus evolution but could also be an effect of campaigns to poison rabbits. Myxomatosis was released in 1950 in south-eastern Australia and poisons including sodium monofluoroacetate (compound 1080) have been used extensively since then in the same region. This hypothesis could be tested by comparing virulence in areas that have and have not used poisons for rabbit control. The issue is further discussed in section 6.4.

Predictions of the model can be compared with the results of a field poisoning of feral pigs using warfarin. The poisoning occurred in south-eastern Australia and was described by McIlroy *et al.* (1989). For the three-week duration of the poisoning, births and natural deaths were assumed to be zero and it was assumed there was no resistance to warfarin and no portion of the population was isolated ($I=0=a=m$). The latter assumption is not always correct, as reported by Hone (1983) and Choquenot *et al.* (1990). The free-feed period was not modelled.

The population density of feral pigs was estimated to be 1.8 pigs/km^2 (McIlroy *et al.*, 1989). The average time to development of signs of poisoning was three days, so $h=1/3=0.333$/day. The average time till death was assumed to be four days and mortality 92% based on results of Hone & Kleba (1984). Hence, the mortality rate (j) was $(-\log_{10}(1-0.92))/4=0.274$/day. It was assumed that pigs did not recover and become susceptible ($k=0$). The rate of change (c) from susceptible to eating poisoned bait was not estimated empirically so a value of 0.0025 km^2/lethal dose/day was used. The units of c are different to that described by Hone (1992a); the change being made here to balance units. The amount of poisoned bait was assumed to be 100 lethal doses/km^2 and it was assumed that the rates of loss of poison from weathering (μ) and feeding saturation (g) were zero.

The equations for the model were then:

$$dS/dt = -0.0025SW \tag{6.33}$$

$$dP/dt = 0.0025SW - 0.333P \tag{6.34}$$

$$dY/dt = 0.333P - 0.274Y \tag{6.35}$$

$$dW/dt = -0.0025NW \tag{6.36}$$

The model predicted a sigmoidal increase in the percentage of pigs that were poisoned and a sigmoidal decline in the percentage of pigs still alive. Poisoned bait was added daily so the equations can be solved to estimate the threshold rate of poison addition needed to minimise the ratio of the number of susceptible to poisoned pigs. The rate from equation 6.25 is:

$$A_c = (cN + \mu)\ (j + k + m)/c$$

$$= (0.0025 \times 1.8 + 0)\ (0.274 + 0 + 0)/0.0025 \tag{6.37}$$

$$= 0.49 \text{ lethal doses/km}^2\text{/day}$$

This is obviously a small amount. The predicted results were compared with the observed results based on the known mortality of 32 pigs that had been fitted with radio-transmitters prior to the poisoning, as described by McIlroy *et al.* (1989). The general trends in mortality were similar (Fig. 6.6) though the pigs appeared to be slower to find the poisoned bait than predicted and then most died very quickly. Choquenot *et al.* (1993) used a version of this poisoning model to model trends in the number of feral pigs trapped at a site in south-eastern Australia.

6.3.2 *Ecological basis of poisoning*

Hone (1986) developed models of poisoning which estimated the probability of an animal dying in a poisoning programme. Four models were developed for different ecological situations: each combination of random and non-random search and random and non-random bait dispersion. A revised version of one of the models (random search and bait dispersion random or non-random) is described here to illustrate the modelling process and the relationship between the process of poisoning and the ecological concepts of predator–prey relationships. The various equations in the models are also represented as graphs to show the ideas. The models show how simple ideas can be built up into a set of complex ideas. Hence, the modelling suggests that other complex ideas or situations may be able to be broken down into a set of simple ideas then reassembled.

The modelling process concentrated on estimating two probabilities. The probabilistic models estimated the probability of dying ($P(\mathrm{D})$) as the product of the probability of the pest animals eating the poisoned bait ($P(\mathrm{E})$) and the probability of dying given that it has eaten the poisoned bait ($P(\mathrm{D/E})$):

$$P(\mathrm{D}) = P(\mathrm{E}) \times P(\mathrm{D/E}) \tag{6.38}$$

Hence, the emphasis in modelling was on the fate of the individual, compared to the fate of the population in the previous section. The probability that an animal eats poisoned bait was assumed to increase with an increase in the number of times an animal finds the bait (t) and to be a function of behavioural interactions between animals that find the bait (i) (Fig. 6.7). In a simple case:

$$P(\mathrm{E}) = \left(\frac{kt}{a+t}\right)^i \tag{6.39}$$

where $t > 1, 0 < k < 1$ and $a > 0$. Here k is the maximum probability or value or $P(\mathrm{E})$ and a is an index of the time required to maximise $P(\mathrm{E})$. Animals showing

Fig. 6.7. The relationship between the probability of an animal eating poisoned bait and the number of times an animal finds poisoned bait as described by equation 6.39. The effect of behavioural interactions is also shown.

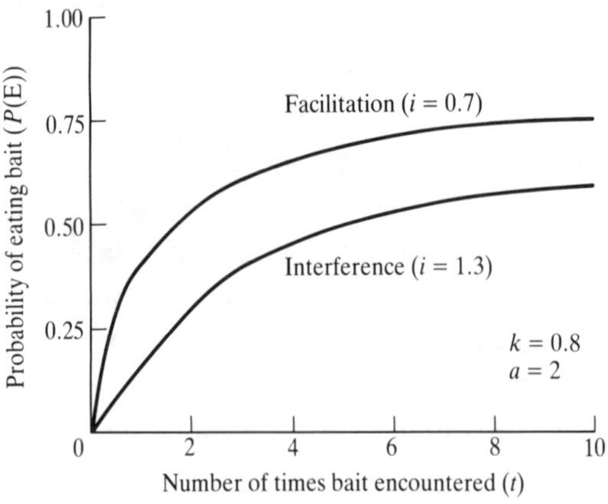

neophobia, such as some rats (Shorten, 1954; Barnett, 1958) and some rabbits (Rowley, 1963; Oliver *et al.*, 1982), will have $k=0$ and hence $P(\mathrm{E})=0$. Barnett & Prakash (1976) described an increase over time in the number of visits by *Rattus norvegicus* to a pile of bait. On successive days more of the bait was eaten. A similar increase was reported for rabbits in southern England (Cowan, Vaughan & Christer, 1987). Trewhella *et al.* (1991) described a temporal increase in bait removal by foxes in Bristol. Bait removal differed significantly, as assessed by Tukey's range test, between habitats, suggesting that the parameter k differs between habitats.

The coefficient i in equation 6.39 equals 1 when animals do not interact; $i>1$ represents behavioural interference between animals such that others decrease the probability and $0<i<1$ represents social facilitation where other animals increase the probability that an animal eats the poisoned bait. Rowley (1958) noted social facilitation of feeding by rabbits. Morgan (1990) reported that captive brushtail possums (*Trichosurus vulpecula*) differed significantly in mean times till they ate toxic pellets compared with times till they ate non-toxic pellets. The mean time for non-toxic baits was less. The toxin was sodium monofluoroacetate (compound 1080). No such difference occurred for carrot baits. Hence, the value of k and/or a may differ for different baits, toxic and non-toxic.

The probability that an animal ingests a lethal dose of poison given that it has eaten the poisoned bait ($P(\mathrm{D/E})$) is assumed to increase at a decreasing rate, with an increase of the dose of poison bait ingested (f) per unit weight of animal (W) (Fig. 6.8). This is based on the classic dose-response relationship

Fig. 6.8. The relationship between the probability of an animal dying given that poisoned bait is eaten and the dose of poisoned bait ingested (f/W) as described by equation 6.40.

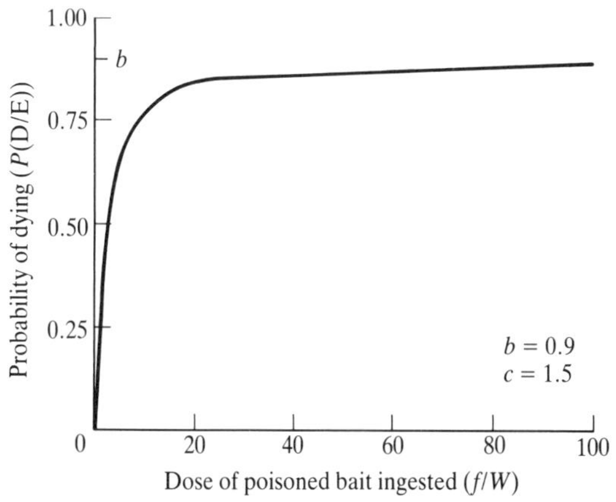

when the dose is expressed on an arithmetic scale (Snyder, 1984). An equation for this is:

$$P(\mathrm{D/E}) = b(f/W) \;/\; ((f/W) + \mathrm{c}) \tag{6.40}$$

$$= b\,f \;/\; (f + cW) \tag{6.41}$$

where $0 < b < 1$, $c > 0$ and b, c and f are estimated empirically. The maximum probability is b. Batcheler (1982) examined this relationship for possums and rabbits though he used a different model, investigating the effects of bait size and toxicity.

The different models suggested by Hone (1986) for random and non-random bait dispersion were based on type III and type II functional responses (respectively) following the laboratory results reported by Real (1979) for mice (*Peromyscus maniculatus*) feeding on wheat. See section 5.4 for a description of types of functional response. However, results of type II functional responses reported for mice (*Onychomys torridus*) feeding on mealworms (Taylor, 1977), kangaroos, sheep and rabbits feeding on vegetation in field enclosures (Short, 1985), bank voles (*Clethrionomys glareolus*) feeding on willow shoots in a laboratory (Lundberg, 1988), penned moose (*Alces alces*) feeding on birch (Lundberg & Danell, 1990), and the theoretical analysis of Abrams (1982) that suggested the functional response may be plastic, indicate that the classification used by Hone (1986) may not have been necessary. Type II and III responses occur but may not always correspond to non-random and random bait dispersion respectively. Also, pest animals usually have an alternative food supply so, as an initial investigation, the two-prey equivalent (Lawton, Beddington & Bonser, 1974) of the random predator equation could be used for modelling the relationship between food availability and intake.

The weight of poison bait eaten (f) is assumed to increase with increases in the weight of poison bait available (B) and the time the poison bait is available (T) (Fig. 6.9):

$$f = B\,(1 - \mathrm{e}^{-d(T - h\,f - g\,A)}) \tag{6.42}$$

where h and g are the handling times per unit of poisoned bait (B) and alternative food (A) respectively and d is the rate of successful search and relates to the efficiency of foraging and the palatability of the food. Rowley (1957) and Cowan *et al.* (1984) described temporal increases in bait consumption by penned rabbits in experiments in Australia and England respectively. The latter study reported that bait consumption dropped slightly after six to seven days. Such a drop has not been included in the model here. The above relationship assumes that the intake of poisoned bait is independent of poison concentration.

An example of violation of an assumption in equation 6.42 was reported by Morgan (1982). He studied bait acceptance by brushtail possums (*Trichosurus vulpecula*) in New Zealand. When the sodium monofluoroacetate (1080) concentration in bait was 0.2%, 75% of possums that survived poisoning did not eat the bait compared with 43% when the concentration of 1080 was 0.1%. Equation 6.42 assumes that intake is independent of poison concentration. The equation could be modified so that the handling time of the poisoned bait, h, is positively related to poison concentration. That is, as poison concentration increases, then handling time increases so intake of poison bait (f) decreases. Alternatively, the coefficient d, which is related to the palatability of the food, may be negatively related to poison concentration.

The handling time of different toxic baits may differ, as Morgan (1990) reported for captive brushtail possums. The median time for eating toxic carrot bait was significantly less than that for toxic pellets. The times reported included the time to find and eat each bait so are different to the handling times in the model.

Substituting for f from equation 6.42 into equation 6.41 gives:

$$P(\mathrm{D/E})=\frac{b\ B\ (1-\mathrm{e}^{-d(T-h\ f-g\ A)})}{B\ (1-\mathrm{e}^{-d(T-h\ f-g\ A)})+cW} \tag{6.43}$$

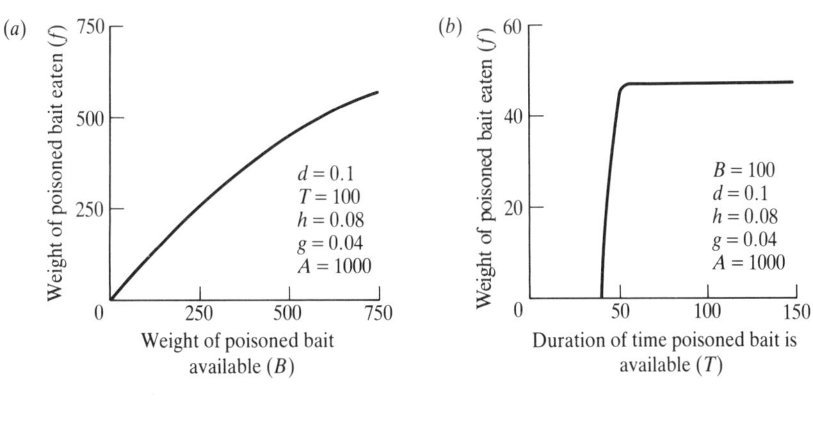

Fig. 6.9. The relationship between the weight of poisoned bait ingested (f) and (a) the weight of poison bait available (B) and (b) the duration of the time poisoned bait is available (T), as described by equation 6.42.

We now have estimates of $P(\mathrm{E})$ and $P(\mathrm{D/E})$ which can be combined to estimate $P(\mathrm{D})$. Combining equation 6.43 with equation 6.39 gives:

$$P(\mathrm{D})=\left(\frac{k\ t}{a+t}\right)^{i}\frac{b\ B\ (1-\mathrm{e}^{-d(T-h\ f-g\ A)})}{B\ (1-\mathrm{e}^{-d(T-h\ f-g\ A)})+c\ W} \tag{6.44}$$

Equation 6.44 indicates that the probability that a pest animal dies from poisoning is a function of 14 factors. As the equation includes two terms, each with divisions, then the value of $P(\mathrm{D})$ will be determined by the relative values of each parameter rather than the absolute values of each. The strategic planning options are defined by equation 6.44 and are summarised in Table 6.3. By strategic timing of poisoning the amount of alternative food available that is eaten (A) and pest weights (W) could be reduced. Such a strategic approach to poisoning rabbits has been suggested for central Australia (Foran

Table 6.3. *Factors that influence the probability of a pest animal dying in a poisoning programme as described in equation 6.44. Control strategies suggested by the model to increase the probability of death are listed and a subjective assessment of the degree of operator control is also shown. The degree of control ranges from direct (complete control) to partial, to limited, to none*

Number	Symbol	Factor	Control strategy	Degree of control
1.	k	Maximum value of $P(\mathrm{E})$	Increase	Limited
2.	t	Times pest finds bait	Increase	Partial
3.	a	Index of time to maximise $P(\mathrm{E})$	Decrease	None
4.	i	Behaviour at bait	Decrease	Partial
5.	b	Maximum value of $P(\mathrm{D/E})$	Increase	Limited
6.	B	Weight of poisoned bait available	Increase	Direct
7.	d	Rate of successful search	Increase	Limited
8.	T	Time bait available	Increase	Direct
9.	f	Weight of poison bait eaten	Increase	Limited
10.	h	Handling time per unit poisoned bait	Decrease	Limited
11.	g	Handling time per unit alternative food	Decrease	None
12.	A	Weight of alternative food eaten	Decrease	Limited
13.	W	Weight of pest	Decrease	Limited
14.	c	Tolerance of poison	Decrease	Limited

Modified from Hone (1986).

et al., 1985). Managing pesticide resistance could decrease the tolerance (c) of the poison by pests. Control operators have direct control over only the time bait is available (T) and the weight of poisoned bait (B). Operators attempt to increase the number of times pest animals find the bait (t) by pre-feeding. Control operators have partial control over the behavioural interactions (i) between pest animals at a bait station, by careful design of bait stations or use of trails. There is limited or no control over the other factors.

The model predicts that increases in the weight of poison bait available (B) will increase the probability of death. Williams *et al.* (1986) reported, in experiments in New Zealand, that as the number of poisoned baits laid increased, the percentage of rabbits killed increased significantly. They fitted a piece-wise (ramp) linear regression to the ascending part of the data. The trend in the data is in agreement with the predicted trend.

The effects of varying the weight of poison bait eaten (f) and the weight of alternative food eaten (A) are not straightforward. By themselves f and A are not directly important but they are important as measures of the time involved in eating. Hence, if A is large then the time involved in eating ($g\ A$) is increased so the weight of poison bait eaten (f) will decrease.

The conclusions and strategic options described by this model are very similar to those of the models given by Hone (1986). The model above, and the model of the dynamics of poisoning, are based on pen and field data. There is a need, however, for further testing of these models of poisoning, and examination of their strategic application.

6.4 Biological control using pathogens

The statistical analysis of biological control was discussed in section 3.11.

Anderson (1982a) examined biological control of pests, mainly insects, using models of disease dynamics. Similar models were described in section 5.6. A slightly different model is described here, following Anderson (1982a), by modelling pathogens that can survive for substantial periods of time in the environment of the host. This does not mean that only such pathogens can be biological control agents, however.

To be useful as a biological control agent a pathogen must initially establish. That occurs when the basic reproductive rate (R) of the infection is greater than, or equal to, 1.0, though if $R=1$ the pathogen will not be very effective as the number of infected hosts will not increase. For the pathogen to regulate the pest population the extra per capita mortality from the disease (α) must exceed the intrinsic rate of increase of the host population (r_m). If the pathogen reduces the birth rate of the host (a) to the level $a(1-f)$ where f is the proportion of infected hosts unable to breed, and the host population increases exponentially in the absence of the disease, then the condition for regulation is:

$$\alpha > a(1-f)-b \tag{6.45}$$

where b is the per capita natural mortality rate. If infected hosts never reproduce ($f=1$) then the pathogen will always regulate the host as $a(1-f)-b=-b$. The host population simply dies out.

Anderson (1982a) listed features of the pathogen–host interaction that favour regulation of a host population by a pathogen (Table 6.4). Hence, selection of pathogens for biological control solely on the basis of high pathogenicity (α) is not sufficient.

Biological control was modelled by introducing the pathogen at a certain rate (A) into the environment (Fig. 6.10). Hence, the basic model incorporated infection of susceptibles by contact with free-living infective stages (W) so the rate of change of susceptibles was proportional to the product of the number of susceptibles (X) and the level of infective stages (W), i.e. βXW where the proportionality constant is the transmission coefficient (β). Anderson (1982a) showed that the pathogen will push the host to extinction if the rate of introduction of pathogens (A) exceeds a critical threshold level (A_c). To estimate the threshold level requires estimates of key disease and population parameters. The threshold, A_c, may be difficult to estimate because the transmission coefficient (β) is hard to estimate. Methods for estimating the transmission coefficient were listed in Table 5.3.

The effects of pathogens as biological control agents can be complicated by simultaneous use of poisons for pest control (Carpenter, 1981; Anderson, 1982a). The poisons may lower average pest abundance but generate oscillations in pest abundance such that occasionally the pests are more

Table 6.4. *Factors that favour the regulation of a host population by a pathogen, as predicted by disease models, and the corresponding parameters in disease models*

Number	Factors
1.	High disease-induced mortality rate (α) compared with natural mortality rate (b)
2.	Disease mortality rate (α) increases with host density
3.	Reduction in host birth rate (a)
4.	Short latent period ($1/\sigma$)
5.	Production of large numbers of long-lived infective stages (W)

After Anderson (1982a).

abundant than when no poison was used. In contrast, if the control method reduced the birth rate of the pest then regulation of the pest by the pathogen was not influenced. The model of Carpenter (1981) differed from that of Anderson (1982a) as pathogens were directly transmitted from infectives to susceptibles, not via a free-living stage. Anderson (1982a) cautioned that more research was needed on the combined use of chemicals and pathogens for pest control to understand the consequences. This issue was briefly discussed in section 6.3.1 for rabbit control using myxomatosis and sodium monofluoroacetate simultaneously.

McCallum & Singleton (1989) modelled the use of a parasite, *Capillaria hepatica*, as a biological control agent of the house mouse. The models suggested that the parasite would not persist in mouse populations so successful biological control would need to be based on tactical releases of the parasite. The threshold rate of parasite release (A_c) was not estimated.

6.5 Shooting

Statistical analysis of shooting was discussed in section 3.12.

Shooting is a simple predator–prey system where a human is the predator and the pest is the prey (Hone, 1990). This analogy suggests that the modelling of the process and dynamics of shooting could be similar to the modelling of predation or foraging generally. Smith *et al.* (1986) modelled the response of

Fig. 6.10. Flow chart of the movement of animals through different stages of infection when infected with a pathogen as a biological control agent that has a long-lived stage (W) in the environment of the host. Births and natural deaths are not shown.

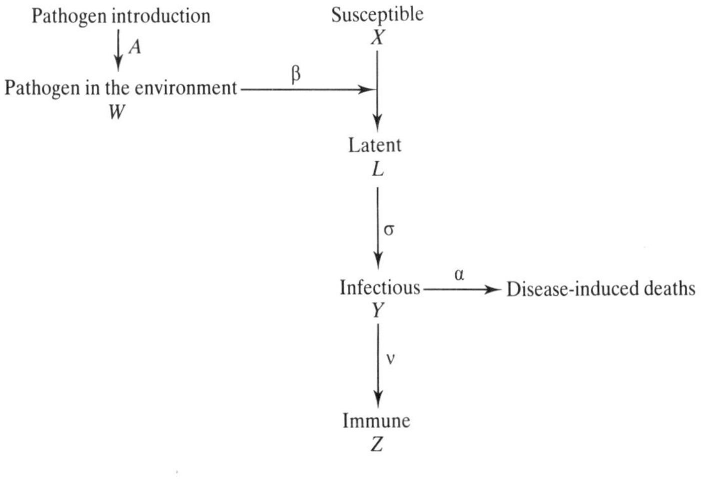

coyote populations to shooting as control but did not examine the internal dynamics of shooting.

6.5.1 *Models*

Predator–prey interactions have been modelled extensively with key components being the functional and numerical responses (Holling, 1959; Krebs, 1985; Begon *et al.*, 1990), as described in section 5.3.1. The functional response describes the change in the number of prey eaten relative to changes in prey density and duration of hunting. The numerical response describes the rate of increase of the predator population relative to the abundance of prey.

The functional response relationship has been represented in many ways, including as a linear model (Nicholson, 1933), by the disc equation (Holling, 1959), by the random predator equation (Rogers, 1972; Lawton *et al.*, 1974), as a general model (Fujii, Holling & Mace, 1986) and as a power function (Tome, 1988). Each of these models assumes a multiplicative relationship between initial density (N) and the duration of hunting (T). The effects of time and initial density could each act in a linear or curvilinear manner. Possible density relationships are shown in Fig. 6.1*b* and *c*. The early Nicholson model assumed linear effects of time and density, the disc equation assumed linear effects of time and curvilinear effects of density, the random predator and general models assumed curvilinear effects of both variables, and the power function assumed linear effects of density and curvilinear effects of time.

As density increases the number of pests shot should increase unless shooters get to shooting saturation where more animals are seen than can be shot and hence the effect of initial density should be curvilinear. If this is true then the use of shooting kill statistics (kills/h) as a linear index of density, as used by O'Brien (1985) for feral pigs, may not be appropriate. If the kill rate is low the various models will be indistinguishable. Murton *et al.* (1974) reported a negative correlation ($r = -0.72$, df$=53$, $P<0.001$) between the percentage of wood-pigeon shot (y) and the common logarithm of the number at risk (x). The negative relationship was attributed to the short time available for a shooter to reload. Hence, the shooter was at saturation at high pigeon numbers.

Increasing the duration of shooting should increase the number of animals shot. The increase could be at a decreasing rate as animals become harder to find. Hence, the effect of time could be curvilinear. Stephens & Krebs (1986) described hypothetical functions of a predator with increasing time in a patch containing prey. Each function (relationship) satisfied two assumptions: (i) the kill was zero when time in the patch was zero, (ii) the kill function was initially increasing. Four functions which satisfy these assumptions are shown in Fig. 6.11.

The random predator equation describes predation when the prey population is depleted by the killing, compared with the disc and general equations which assume that prey density is constant (Fujii *et al.*, 1986; Juliano & Williams, 1987). The latter two equations are of little use in modelling vertebrate pest control except when it has little effect on pest abundance. The predator–prey models have some other limitations, as described by Stephens & Krebs (1986). For example, they may not account for hunter satiation (for example, a quota) or tiredness. The models do not always allow foragers to use information on prey abundance or distribution (such as by communicating by radio), and many only allow sequential and not simultaneous encounters with prey. A detailed listing, and discussion, of the assumptions and predictions of some models of optimal foraging was given by Krebs & McCleery (1984).

The differing susceptibility of prey is an important source of population stability in predator–prey relationships (Murdoch & Oaten, 1975). Also identified (with analogies to pest control inserted here) were prey refuges (for example, vegetative cover), limits to predator dispersal (tiredness of shooters or fuel supply for a vehicle or helicopter), and switching by predators (shooters

Fig. 6.11. Some possible relationships between the net energy gained by a predator and the duration of time the predator is in a patch containing prey. Graphs (*a*) to (*c*) satisfy the criteria of the marginal value theorem (Charnov, 1976) and graph (*d*) corresponds to systematic search by the predator and random distribution of the prey. (After Stephens & Krebs, 1986.)

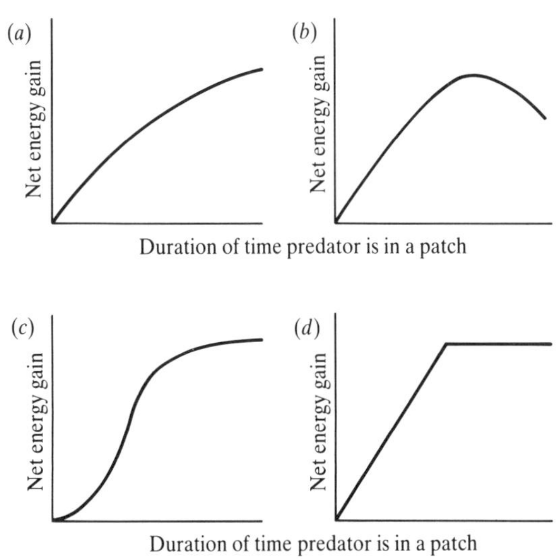

changing targets to shoot other pests). The implications for vertebrate pest control are that each of those sources of stability is a source of pest survival and maybe higher pest damage.

The models of predator–prey relationships can also be used to study other aspects of shooting. These are now described.

6.5.2 *Where to shoot from*

The theory of central place foraging (Stephens & Krebs, 1986) could apply to shooters travelling from a set site, hunting and then returning to that site. If the hunter returns after one kill, it is a single-prey loader, but if the hunter returns after several prey have been killed it is a multiple-prey loader (Orions & Pearson, 1979; Stephens & Krebs, 1986). If hunters are not restricted to a single central set site but use one of a limited number of sites then they can be called multiple central place foragers (Chapman, Chapman & McLaughlin, 1989). Such a foraging strategy should be advantageous when resources (pests) are low in abundance. If resources (pests) are very scarce neither may be efficient (Chapman *et al.*, 1989).

6.5.3 *How long to search*

The marginal value theorem for prey capture stated that a predator should stay in a patch until the capture rate attained its maximum value (Charnov, 1976), which can be graphed as shown in Fig. 6.12. The model includes the travel time between patches, in contrast to many other models of prey capture. Krebs & McCleery (1984) concluded that subsequent studies

Fig. 6.12. The relationship between prey captures and the duration of hunting by a predator. Prey capture only starts after a predator gets into a patch so the curve does not start at the origin (0,0). The maximum capture rate is shown by the tangent to the curve and occurs at time *T*. (After Charnov, 1976.)

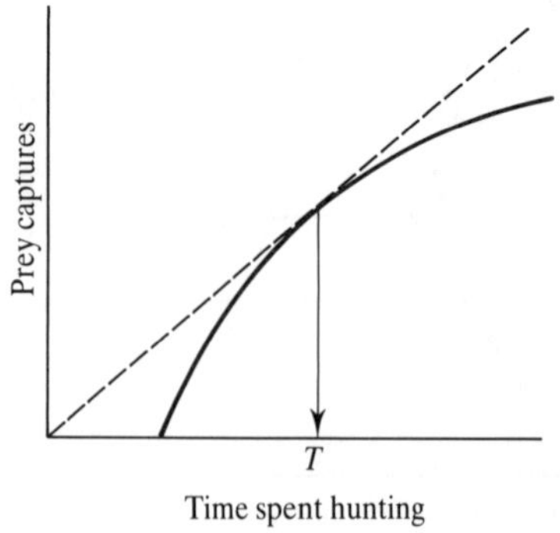

suggested that if a forager (hunter) was acquiring information about food in a patch then it should stay longer than predicted by the marginal value theorem.

The foraging behaviour of birds supported the hypothesis that predators should stay longer in patches when it takes the predator longer to get there (Cowie, 1977). The relationship between travel time and stay time was positive but curvilinear. Krebs & McCleery (1984) describe other such studies. There are other influences on the duration of hunting in an area. Green (1984) examined the stopping rules for predators feeding in patches of prey. The strategy of a predator staying in an area (patch) for a fixed time was the best rule only if prey were randomly distributed. As animals are usually clumped this strategy appears to be of limited value. Hone & Bryant (1981) had suggested this shooting strategy for use in a plan to eradicate feral pigs from an area in Australia. An alternative approach is to stay in an area until a certain time has elapsed since shooting the previous animal. This is the giving-up time (GUT) rule of optimal foraging. Iwasa, Higashi & Yamamura (1981) concluded that the GUT rule was the best strategy to maximise kills when the prey had a clumped distribution. Murton *et al.* (1974) described a study of shooting to control wood-pigeon in a part of England, and noted that shooters left a site after a reasonable time if no birds appeared. The duration of the reasonable time was not specified. The shooters appear to have used a flexible strategy that had features of the giving-up time and the fixed-time strategies. Ridpath & Waithman (1988) used a variation on the GUT strategy in northern Australia by fixing the maximum time required to shoot a buffalo (*Bubalus bubalis*) as 15 minutes. If the duration of shooting (the total time for searching and actually shooting) exceeded 15 minutes then shooting stopped.

6.5.4 *Effects of cover*

The effect on predator–prey systems of hiding behaviour by prey has been examined briefly by Maynard Smith (1974). If the number of prey that have cover is constant the population system tends to stability but not if a constant proportion of prey have cover. Hughes (1979) noted that if a proportion, p, of prey (pests) hide and hence cannot be shot then $N(1-p)$ is the total number of pests that can be shot. Saunders & Bryant (1988) reported such an event when two of six feral pigs with radio-transmitters were never seen by shooters despite intensive searching and shooting from a helicopter. The pigs apparently hid in thick shrubs. Taylor (1984) examined the effect of prey clumping and crypsis on vulnerability to predation. He concluded that prey can benefit from aggregation, above a threshold group size, if the prey do not hide or seek refuge. When prey use crypsis (hide or rely on camouflage) aggregation can be harmful if the probability of a predator successfully attacking a group increases with increasing group size.

The relationship between predator success and habitat cover was examined by Gotceitas & Colgan (1989). They modelled the relationship as reversed sigmoidal (as cover (x) increased, success (y) declined) using logistic regression. Their experiments used fish, which were not pests, and vegetation, and the test fish were prey of predatory fish. The relationship between predator success and cover was significantly different from a linear regression. Hence, predator success was high when cover was low and lower when cover was higher. The fitted model was:

$$y = 1 \;/\; (1 + e^{-a-bx}) \tag{6.46}$$

The coefficients a and b were estimated by regression. The inflection point was equal to $-a/b$, and is the level of cover at which predator success changes most rapidly.

6.5.5 *Analysis of data*

Shooting data can be used to estimate how much time is needed to remove a certain percentage of animals or an individual animal (that is, number removed and initial number known and equations solved for time) and secondly, to obtain estimates of initial numbers when shooting occurs (that is, number removed and time known and equations solved for initial number). Data can also be used to estimate cost-effectiveness of shooting (section 4.2.3) or as part of a cost–benefit analysis (section 4.4).

Hone (1990) reported an example of the first use. Two data sets for shooting feral pigs in Australia were analysed by non-linear least squares regression. For one set the best-fit equation was:

$$n = 4.834\ N\ T^{0.870} \tag{6.47}$$

where n was the number removed (/km^2), N was initial density and T was the duration of shooting (hours/km^2). The equation estimates that to reduce initial density by 90% would take 0.145 hours/km^2 of shooting.

The second data set, from Saunders & Bryant (1988), had a best-fit equation of:

$$n = N(1 - 1.184e^{-9.921T}) \tag{6.48}$$

A 90% reduction of this population would take 0.249 hours/km^2. The estimates were higher (longer times) than those from the regressions that assumed linear relationships of time and number removed (0.138 and 0.194

hours/km^2 respectively). The curvilinear responses of time and number shot (of the type shown in Fig. 6.11a) could be associated with animals becoming harder to find and shoot. Vitale (1989) reported changes, with age and experience, in anti-predator responses of wild rabbits.

Equation 6.47 is an example of the Cobb–Douglas function (expression 8) listed in Table 6.1, except here n is the cumulative number of pests removed not the rate of removal. Equation 6.48 is an example of expression 12 in Table 6.1, slightly modified to include the coefficient (1.184) of the exponential term. In both equations effort (E) is not explicit but equals T.

The estimates of best-fit for the linear and curvilinear regressions reported by Hone (1990) were very similar (all coefficients of determination greater than 0.98), which highlights the difficulties that can occur of interpreting results of simple curve-fitting exercises that have no theoretical background. Such situations can occur in analysis of shooting data when there are no estimates of initial pest abundance. Murton *et al.* (1974) described data from a study of shooting wood-pigeon in a part of England. My analysis, as described in section 3.19.2, of data in their Table 2 shows a highly significant ($P<0.01$) correlation ($r=0.92$, df$=12$) between the number of wood-pigeons shot (y) and shooting hours expended (x). There was no evidence of the relationship being curvilinear. Models of the functional response relationship cannot be used to analyse the data as the initial abundance of the pigeons in each month was not reported.

6.6 Fertility control

Statistical analysis of fertility control was discussed in section 3.13 and the effects of biological control agents that reduce host fertility have been noted in section 6.4. A different approach to modelling of fertility control can also be examined. The actual method of fertility control is not considered but rather the effects of any such control. The aim of fertility control work is to lower pest abundance and hence hopefully lower pest damage. The decline of a pest population occurs when the instantaneous rate of increase (r) is negative. The relationship between the rate of increase and fertility should therefore be examined. Hone (1992b) described two methods for examining the relationship. The first method is described here.

The instantaneous rate of increase is related to the net reproductive rate (R) and the generation interval (T) by:

$$r=(\ln R)\ /\ T \tag{6.49}$$

The net reproductive rate (R) is defined as the number of female young produced by a female during its lifetime (May, 1981a; Crawley, 1986), and the generation interval (T) as the average age of females when giving birth to young (Millar & Zammuto, 1983), or the mean age of mothers of all newborn females in a population with a stable age distribution (Caughley, 1980). Caughley (1980) and Crawley (1986) cite the relationship as exact though Laughlin (1965), May (1976), May (1981a), Crawley (1983) and Krebs (1985) cite the relationship as approximate. Any bias of the equation was described by May (1981a) as being low when R is approximately one or the generation interval is not too variable.

The relationship between the measure of fertility (R) and rate of increase (r) is curvilinear (Fig. 6.13). The population is stable ($r=0$) when $R=1$, as $\ln 1=0$. The reproductive rate and generation interval of muskrat (*Ondatra zibethicus*) in Europe were estimated by van den Bosch *et al.* (1990) as 3.1 (dimensionless) and 1.41 (years) respectively. This equates to:

$$r = (\ln 3.1) \,/\, 1.41 = 0.80 \text{ /year} \tag{6.50}$$

If reproductive rate dropped to 2.1 then $r=0.53$/year, and if $R=1.1$ then $r=0.07$/year, and if $R=0.1$ then $r=-1.63$/year. Hence, the unit reductions in fertility cause drops of 0.27, 0.46 and 1.70 respectively in the rates of increase. The rate of increase declines at an increasing rate. Van den Bosch *et al.* (1992)

Fig. 6.13. The relationship between the instantaneous rate of increase of a vertebrate pest population and the net reproductive rate as described by equation 6.49. The effects of increasing (––––) and deceasing (······) the generation interval (T) are also shown.

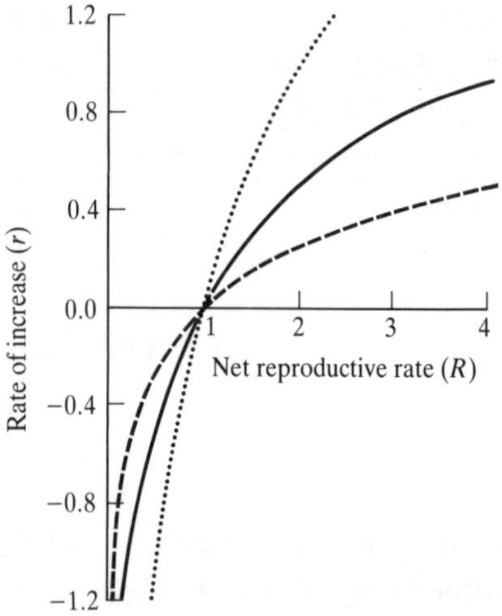

give worked examples, for five species of vertebrates, of estimates of net reproductive rate and generation interval.

The effects of fertility control may not be so clear if the generation interval is not constant. The outcome will be determined by the ratio of the two changes not by the absolute amounts of each change. The generation interval will vary with some changes in fertility. If females have fewer births later in life then the generation interval will be reduced. If females have fewer births early in their lives, then the generation interval will be longer. If the females have fewer births during mid-life then the generation interval will be unchanged. This is necessarily simplified because the survivorship pattern of females is unlikely to be linear. The effect of varying the generation interval is illustrated in Fig. 6.13, which shows a reversal in effect when net reproductive rate decreases below unity. The response is more complex than that described by Crawley (1983) of a decrease in generation interval causing an increase in the rate of increase. There are biological limits to how far generational interval can vary. The lower limit is puberty. If generation interval is less than puberty then by definition there are no young produced. An upper limit is life expectancy, as animals can not produce young after they have died.

There are two implications of the relationship that are of practical significance. Firstly, the response of a vertebrate pest population will vary depending on the initial level of fertility. Secondly, the greatest response to fertility control occurs in populations that are already declining, that is, have a negative rate of increase. The results are similar to those reported by Garrott (1991) who modelled, by computer simulation of an age-structured model, the effects of fertility control on populations of feral horses. Hone (1992b) further examined, using Lotka's equation, the topic of rate of increase and fertility control and reported similar results.

The model of poisoning described in section 6.3.1 could be used to predict the effects of fertility control on pest dynamics. The model would be more relevant for control by use of a chemosterilant.

An alternative analysis of fertility control can be described. A pest population has a certain level of fertility. The net reproductive rate, R, is a useful measure of such fertility. Fundamentally, fertility control is about reducing fertility, often so that pest abundance or pest damage declines. If pest abundance is reduced so that abundance is then stabilised, the net reproductive rate is then unity. Pest fertility, as a proportion of initial fertility is then $1/R$, the level of fertility control is therefore $1-(1/R)$. This is actually a threshold, p, as fertility control greater than this level will ensure that abundance declines and hence the exponential rate of increase is negative. Fertility control to less than the threshold will maintain an increasing population. The equation for the threshold is:

$$p = 1 - (1/R) \tag{6.51}$$

and is illustrated in Fig. 6.14. Specific data can be used to show the key points. The net reproductive rate of collared doves (*Streptopelia decaocta*) during their spread in Europe was estimated as 1.33 (van den Bosch *et al.*, 1992). Hence, the level of fertility control needed to stabilise abundance is $1-(1/1.33)=0.24$. If the net reproductive rate dropped by more than 24% then abundance declines. In contrast, the percentage reduction for muskrat, with a net reproductive rate of 3.1, is $1-(1/3.1)=68\%$.

Other studies have modelled fertility control. Barlow (1991b) examined the topic for control of brushtail possums and the prevalence of tuberculosis in New Zealand. Deterministic compartment models of disease and possum dynamics were used similar to those described in section 5.6. Caughley, Pech & Grice (1992) used a probability model to estimate the effects of fertility control on pests. They examined the effects of three forms of social dominance, two effects of sterilisation on dominance and four modes of transmission of the sterility agent. As the percentage of females sterilised increased, then the average number of litters per group declined, as expected. However when sterilisation stopped a dominant female from breeding but allowed

Fig. 6.14. The predicted relationship between the proportional reduction in net reproductive rate needed to stabilise population growth and net reproductive rate, as described by equation 6.51. Note that the relationship starts at (1,0) as a population with a net reproductive rate of 1 is stable and hence has a rate of increase of zero. Examples are shown for collared dove (open circle) and muskrat (closed circle), with data from van den Bosch *et al.* (1992).

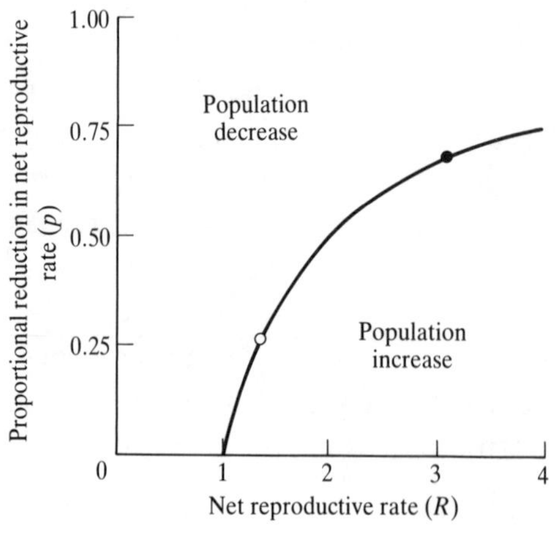

subordinates to breed, it was predicted that sterilisation could allow an increase, within limits, in average litters per group. Hence, fertility control could be counter-productive. The modelling analysis of range expansion by wildlife (van den Bosch *et al.*, 1992) could be used to estimate the effect of fertility control on such range expansion. This could be done because the velocity of spread is linked mathematically to net reproductive rate using equation 6.49.

6.7 Predation control

Statistical analysis of predation of livestock was discussed in section 2.4 and analysis of predation control in section 3.15. Economic analysis was discussed in section 4.4 and modelling of predation in section 5.5.

There is scope for development and testing of models of predation control. Emphasis has been on models of coyote control rather than predation control. A graphical model of the effect of intensive control on seasonal abundance of coyotes in parts of northern America was described by Knowlton (1972). The model incorporated seasonal breeding, dispersal and seasonal changes in the effectiveness of control. Smith *et al.* (1986) also described a model of coyote control, as discussed in section 5.5.

6.8 Control of infectious diseases

Statistical analysis of damage caused by infectious diseases was discussed in section 2.5 and analysis of disease control in section 3.16. Economic analysis was discussed in section 4.5 and modelling of host and disease dynamics in section 5.6.

6.8.1 *Vaccination*

The spread of an infectious disease in wildlife may be stopped by creating a buffer zone. Three options have been investigated: removal of all susceptible hosts, reduction of a population to below a threshold host density, or conversion of susceptible hosts to immune hosts by vaccination. Study of rabies in European foxes suggests that the latter may be more effective, though more study is required (Bacon & Macdonald, 1980; Blancou *et al.*, 1986). The disease model described earlier (section 5.6) can be used to estimate the proportion of the host population that needs to be vaccinated to achieve the effect of disease eradication. Vaccination *per se* is not a method of vertebrate pest control, but can be used to convert a vertebrate pest to a harmless wild animal.

If a proportion of the host population (p) is vaccinated shortly after birth, then the number of secondary infections, the basic reproductive rate of the disease (R), will be reduced (Anderson & May, 1982b):

$$R_a = R\ (1-p) \tag{6.52}$$

where the actual reproductive rate is R_a. As it is necessary for R_a to be greater than one for a disease to establish then:

$$R\ (1-p) > 1 \tag{6.53}$$

Therefore:

$$p > 1-(1/R) \tag{6.54}$$

This equation indicates that as R increases then the proportion of the host population that needs to be vaccinated (p) must increase though not in a linear manner (Fig. 6.15). For example if $R=2$, $p=0.5$ and if $R=4$, then the proportion of animals to be vaccinated is $p=1-(1/4)=0.75$. This statement is valid when comparing different host populations with the same disease or the same host population with different diseases.

Since $R=X/K_T$ (equation 5.45) and if $X=K$ then, $R=K/K_T$ where K is carrying capacity and K_T is the threshold animal density, then equation 6.54 can be rearranged to give:

$$p > 1-(K_T\ /\ K) \tag{6.55}$$

If $K<K_T$ then $p=0$. Anderson *et al.* (1981) reported parameter estimates to

Fig. 6.15. The relationship between the proportion of a host population that needs to be vaccinated (p) and the basic reproductive rate of the disease (R) as described by equation 6.54. Note that the curve does not start at the origin, but at (1,0) as R must be greater than 1 for disease establishment. (After Anderson, 1984.)

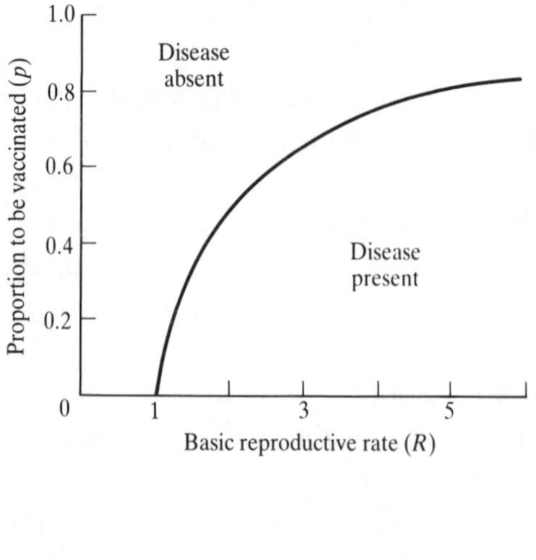

allow calculation of the proportion of a fox population that needs to be vaccinated given that rabies occurs. If the carrying capacity (K) was 2 foxes/km^2, and the threshold density (K_T) was 1.0 fox/km^2, then using equation 6.55:

$$p = 1 - (1/2) = 0.5$$

However, in good habitats, where carrying capacity is higher, the proportion to be vaccinated increases, for example, if $K = 15/\text{km}^2$ then p increases to 0.93. Field studies in urban Bristol showed that up to 35% of adult foxes ate bait of a sort that could be used to distribute an oral vaccine (Trewhella *et al.*, 1991). Higher percentages of foxes have been dosed with bait or vaccine in similar studies in Europe or northern America (Trewhella *et al.*, 1991).

Similar calculations can be done for rabies in raccoons, using data in Coyne *et al.* (1989), where the threshold density (K_T) was 3.0/km^2 and carrying capacity (K) was 12.69/km^2. Equation 6.55 gives:

$$p > 1 - (3.0/12.69)$$

That is, $p > 0.76$.

Coyne *et al.* (1989) used a slightly different equation:

$$v > a((K/K_T) - 1) \qquad (6.56)$$

where v is the instantaneous vaccination rate and a is the instantaneous birth rate (1.34/year). That estimated that $v = 4.33$/raccoon/year, so the proportion to be vaccinated (p') was $p' = 1 - e^{-4.33} = 0.99$.

Note that Coyne *et al.* (1989) estimated the proportion of raccoons to be vaccinated assuming that rabies does not currently occur in the raccoon population. Such vaccination would reduce the population density of susceptible raccoons to less than the critical threshold. Hence, rabies would then not establish. The other estimate here ($p = 0.76$) is of the proportion of raccoons to be vaccinated given that rabies is currently in the population, though in the simplest interpretation there is only one infective ($Y = 1$) and the number of susceptibles (X) is virtually identical to carrying capacity (K). Both proportions are high, especially 0.99, suggesting that rabies may not be controlled easily by vaccination.

Anderson & May (1982b) showed that equation 6.54 could be changed to give:

$$p > (1 + (V/L))/(1 + (A/L)) \qquad (6.57)$$

where V is the average age at which individuals are vaccinated, A is the average age at first infection and L is the average life expectancy. Since the proportion of the population vaccinated (p) cannot exceed one, then for disease eradication $V < A$. That is, the average age at vaccination must be less than the average age at infection. Hence, diseases with high values of A will be easier to control than diseases with low values of A, and control will be easier if V is low rather than high. Essentially this says, vaccinating individuals that are already infected is not very efficient. Anderson & May (1982b, 1985) provide estimates of V, A and L for various human diseases, but I know of no estimates of V for diseases of wildlife. Cheeseman *et al.* (1989) provided some estimates of age at first infection for badgers and bovine tuberculosis.

The eradication of an exotic pathogen in a heterogeneous population was examined by modelling by Travis & Lenhart (1987). They concluded that eradication using vaccination should be based on isolation of infectives to reduce the transmission rate of the disease between infectives and susceptibles. That same conclusion will apply whether vaccination is used or not, as the primary requirement is isolation of infectives. That could be accomplished, in theory, by containment by fences.

The estimates of the proportion of the population to vaccinate are based on the assumption of homogeneous mixing for disease transmission. Travis & Lenhart (1987) concluded on the basis of modelling that the effect of non-homogeneous mixing was for equation 6.54 to underestimate p, the proportion of the population to vaccinate. Hence, as a cautious policy any estimates of p should be viewed as conservative and as a minimum level of protection to be attained.

Barlow (1991b) modelled the effect of vaccination on the prevalence of bovine tuberculosis in brushtail possums in New Zealand. A deterministic compartment model of tuberculosis dynamics was used and predicted that the required rate of vaccination was highly sensitive to assumptions of density-dependent mortality or reproduction. Vaccination did not have any substantial effect on possum abundance, which was interpreted as a disadvantage as possums are pests in New Zealand.

The model of poisoning described in section 6.3.1 could also be used to describe changes in the vaccination status of pests and estimate the threshold rate of vaccination needed to control an infectious disease.

6.8.2 *Culling for disease control or eradication*

The control of an infectious disease by culling is essentially an attempt to reduce host abundance or population density to reduce the number of secondary infections (the basic reproductive rate, R) to less than one. This strategy follows from the positive relationship between secondary infections and pest abundance (equation 5.46). The reduction may be total or to less than the threshold abundance, or density, of susceptible hosts. For disease eradication, the abundance of susceptibles must be held below the threshold.

Culling may be very specific by removing susceptibles or generalised by simply lowering abundance of hosts irrespective of their disease status. The latter is much easier to do in the field. For diseases spread by vectors, an additional strategy is to lower the abundance of vectors, especially infected vectors. Again, there is usually little chance of selecting out the infected vectors from the uninfected, so vectors generally are reduced in abundance.

One approach to modelling of disease control by culling is to include an extra source of mortality in the model. If the added per capita mortality rate (c) is applied to all segments of the host population, when births (a) and natural deaths (b) are included, then equations 5.36 to 5.39 of Chapter 5 become:

$$dX/dt = -\beta XY + \gamma Z + aN - bX - cX \qquad (6.58)$$

$$dI/dt = \beta XY - \sigma I - bI - cI \qquad (6.59)$$

$$dY/dt = \sigma I - \alpha Y - vY - bY - cY \qquad (6.60)$$

$$dZ/dt = vY - \gamma Z - bZ - cZ \qquad (6.61)$$

The threshold host abundance (K_T) is then:

$$K_T = (\sigma + b + c)\,(\alpha + v + b + c)\,/\,\beta\sigma \qquad (6.62)$$

The effect of the extra mortality is to increase the threshold host abundance making it harder for the pathogen to survive in the host population (Carpenter, 1981). This is discussed further in section 6.3.1 for myxomatosis and poisons.

Anderson *et al.* (1981) and Anderson (1982c) modelled control of rabies in European foxes and reported two methods of culling foxes: constant quota and at a constant per capita rate. If foxes are removed using a constant quota, n, then it can be modelled, if logistic growth occurs in the absence of rabies, as in equation 6.8:

$$dN/dt = rN(1 - (N/K)) - n \qquad (6.63)$$

For the culling to stop population growth, at which time $dN/dt = 0$, when the threshold density is greater than half the carrying capacity ($K_T > K/2$), then:

$$n > rN(1 - (N/K)) \qquad (6.64)$$

When $N = K_T$ then:

$$n > rK_T(1 - (K_T/K)) \qquad (6.65)$$

If $K_T = 1.0$ fox/km^2 then:

$$n > r(1 - (1/K)) \qquad (6.66)$$

The carrying capacity (K) of 2 foxes/km^2 represents the upper limit for rabies control to occur, as then $K_T = K/2$. When $K_T < K/2$, then culling must exceed the maximum rate of net recruitment ($rK/4$), which occurs at $K/2$ foxes, and if so culling will push the fox population and rabies to extinction.

Alternatively, a constant per capita culling rate (c) could be used. The threshold rate of such culling is given by equation 6.9:

$$dN/dt = rN(1-(N/K)) - cN \tag{6.67}$$

At equilibrium, $dN/dt = 0$, and after rearranging then:

$$cN = rN(1-(N/K))$$

Which can be rearranged further, by cancelling N, and if $N = K_T$ then:

$$c = r(1-(K_T/K)) \tag{6.68}$$

Since the basic reproductive rate (R) of rabies is equal to K/K_T, as discussed in section 6.8.1, then $1/R = K_T/K$. Hence:

$$c = r(1-(1/R))$$

Hence the result stated by Anderson *et al.* (1981) and Anderson (1982c) that the culling rate must exceed a certain rate:

$$c/r > 1-(1/R) \tag{6.69}$$

If carrying capacity (K) equals 2.0 foxes/km^2 and the threshold density (K_T) equals 1.0 fox/km^2 then $R = 2.0$, and if the intrinsic rate of increase of a fox population is 0.5/year then $c > 0.5(1-(1/2))$, that is $c > 0.25$ /year. This instantaneous rate can be converted to a finite rate (A) as $A = 1-e^{-c} = 1-e^{-0.25} = 0.22$ /year. Hence 22% of the fox population needs to be culled annually to control rabies.

Coyne *et al.* (1989) assumed a constant per capita culling rate (c) to estimate the culling rate needed to control rabies in raccoons in the eastern U.S.A. Given parameter values of a rate of increase (r) of 0.504/year, the threshold raccoon density of 3.0/km^2, raccoon carrying capacity (K) of 12.69/km^2 then $c > 0.504(1-(3.0/12.69))$. That is, the culling rate is, $c > 0.385$. Converting the instantaneous culling rate (0.385/y) to a finite rate gives the cull $= 1-e^{-0.385} = 0.32$. Hence over 32% of the raccoon population must be killed each year to keep population density below the threshold for eradication of rabies.

The effect of culling for control or eradication of bovine tuberculosis in brushtail possums in New Zealand was modelled by Barlow (1991b) using deterministic compartment models. Culling was predicted to eliminate tuberculosis and gave a more rapid decline in abundance of tuberculosis-infected possums than either fertility control or vaccination. In the model by Barlow (1993) the location of control was decided after sampling for diseased possums. A version of equation 2.4 was used to determine how many possums to sample to detect tuberculosis.

6.8.3 *Barriers for disease control*

The above discussions focus on disease control in a population. A spatial component to disease control, that of stopping disease spread, can be studied by modelling spread as described in section 5.3.3 and modelling a barrier. The initial aim of the modelling is to estimate the width of the barrier, that is the distance (in kilometres) between where the disease occurs uncontrolled and where the disease does not occur. Note that what Kallen *et al.* (1985) termed a barrier is different to that referred to by Taylor & Martin (1987) in their discussion of fences in Zimbabwe. The latter study described a fence as a barrier and the strip of land within which wildlife were culled as a cordon sanitaire. The cordon sanitaire of Taylor & Martin (1987) is the barrier of Kallen *et al.* (1985).

Kallen *et al.* (1985) estimated the width of a barrier needed to stop the spread of rabies in foxes. The estimate assumes that fox control occurs within the barrier before rabies arrives, and that fox control reduced fox density by a specified percentage. The width (l) (in kilometres) was estimated by:

$$l = L\ (D/\beta X)^{0.5} \tag{6.70}$$

where L is the width in dimensionless terms, D is the diffusion coefficient, β is the transmission coefficient and X is the initial population density of susceptible foxes. If L was 8, the diffusion coefficient D was 60 km^2/year, and the product of the transmission coefficient and the initial population density of susceptibles (βX) was 20/year, then:

$$l = 8 \times (60/20)^{0.5} = 13.9 \text{ km}$$

The equation can also be expressed (Dobson & May, 1986) as:

$$l = L\ (D/\alpha R)^{0.5} \tag{6.71}$$

where α is the fox mortality rate caused by rabies (10/year) and R is the basic reproductive rate of the disease (2). Hence:

$$l = 8 \times (60/(10 \times 2))^{0.5} = 13.9 \text{ km}$$

The estimate is less than the empirical barrier of 20 km which was reported as being successful in Denmark in stopping the spread of rabies (Kallen *et al.*, 1985), though Macdonald (1980) was not convinced, as noted in section 3.16. Murray *et al.* (1986) estimated that the break width (l) could be between 10 and 25 km depending on assumptions in their model, especially the value of the diffusion coefficient (D). If some foxes develop immunity to rabies the break width is reduced (Murray & Seward, 1992).

Mollison (1986) considered deterministic models of barriers inadequate as they could not describe the individual nature of disease spread, and the process of killing hosts in the barrier may change the host behaviour sufficiently to change the diseases' transmission rates. In theory, deterministic models would require removal of all susceptible hosts to stop spread of a disease. Alternatively, the susceptible hosts could be converted to immune hosts by vaccination, so that the susceptible population was below the threshold density for disease establishment. Barlow (1993) predicted that an area of intensive possum control, which he called a buffer, was of little use in stopping spread of bovine tuberculosis in possums. It was necessary for the width to be greater than the maximum juvenile dispersal distance of about 12 kilometres.

6.9 Rodent damage control

Statistical analysis of rodent damage was described in section 2.6 and analysis of damage control in section 3.17. Economic analysis was discussed in section 4.6 and modelling in section 5.7.

Gosling *et al.* (1983) modelled the effects of trapping and climate on coypu abundance in south-eastern England. Cold winters reduced abundance but unless trapping effort was high, coypu increased after cold winters back towards their initial abundance. A model of rodent control that combined spatial and temporal aspects of control concluded that treatments to increase local eradication rates should be applied uniformly over a region but varying in time as much as possible (Stenseth, 1977). It was predicted that eradication efforts should not be correlated with pest abundance. In contrast, the eradication of coypu in England used a diametrically opposed strategy, with trapping effort being allocated proportional to population size (Gosling & Baker, 1987). The success of the coypu programme suggests a deficiency in the modelling work. Another study of modelling rodent control (Montague *et al.*, 1990) was discussed at the start of this chapter.

There is scope for further modelling of the effect of rodent control on the level and variability in rodent damage. This is particularly so given the results of Tobin *et al.* (1993) who reported that a significant reduction in rat abundance may not correspond to a significant reduction in rat damage.

6.10 Control of bird strikes on aircraft

Statistical analysis of bird strikes was discussed in section 2.7 and analysis of control in section 3.18. Economic analysis was briefly discussed in section 4.7 and modelling of strikes in section 5.8.

A model of bird strikes was used to make suggestions on control of such strikes (Major *et al.*, 1986). The model estimated the time period between bird strikes, so estimated impact rates on turbine rotor blades. Such data could assist in better engine design.

6.11 Control of bird damage to crops

Statistical analysis of bird damage was discussed in section 2.8 and analysis of control in section 3.19. Economic analysis was discussed in section 4.8 and modelling in section 5.9.

A model of the dynamics and control of the finch, *Quelea quelea*, described the exponential rate of increase of a hypothetical population in Africa as a function of juvenile and adult survival and breeding rates of adult females as estimated by a life table analysis (Jones, 1989). The rate of increase of the population declined at an accelerating rate when the cull was high. The effect of culling on the rate of increase was less if the population compensated for culling, for example by having a higher survival rate.

6.12 Rabbit damage control

Statistical analyis of rabbit damage was discussed in section 2.9 and analysis of control in section 3.20. Economic analysis was discussed in section 4.9 and modelling dynamics in section 5.10.

A graphical model of rabbit control in central Australia suggested that a strategic time for control was after a prolonged drought, heavy predation or irregular outbreaks of myxomatosis (Foran *et al.*, 1985). Such timing of control was expected to prevent a sudden increase in rabbit abundance in good seasons. The model was not tested. Models of the effects of myxomatosis on rabbit abundance have been described earlier in section 5.6. There is scope for modelling the effects of other control methods on rabbit abundance and rabbit damage.

6.13 Non-target effects of control

Aspects of statistical analysis of non-target effects of control were discussed in section 3.22. This section focusses on modelling.

Spurr (1979) estimated, qualitatively, the risk to some bird species in New Zealand of not recovering after a reduction in abundance. The reduction could occur after accidental poisoning with sodium monofluoroacetate (compound

1080). The likelihood of population recovery was assumed to be determined by the reproductive and dispersal capacities of birds. Species such as the brown kiwi (*Apteryx australis*) with a high risk of non-recovery had low reproductive and dispersal capacities. The species in the low-risk category such as the kingfisher (*Halcyon sancta*) had both high reproductive and dispersal capacities.

Grant, Fraser & Isakson (1984) studied the effects of four hypothetical control programs, aimed at coyotes, on great horned owl (*Bubo virginianus*) populations. The dynamics of the coyote and owl populations were modelled by use of Leslie matrices which described the age structure of each population and associated birth and death rates. The four control programmes were no coyote control or control every year, or every second or fifth year. The control used meat baits containing sodium monofluoroacetate (compound 1080).

When owl survival rates dropped 5% from poisoning in each year of coyote control there was very little effect on predicted abundance of the owls. When owl survival rates dropped 50% in each control year there were substantial reductions in predicted owl abundance. The recovery times from such reductions were estimated to be more than two decades because of the low birth rates of the owls. As expected the effects were lower when coyote control was every fifth year compared with every year. Validation of the models' predictions was not possible with field data. Instead, questionnaires were sent to experts on great horned owls. Grant *et al.* (1984) stated that the two experts who replied agreed that the model was reasonable, but both commented on the apparent limited amount of data on which to base their evaluations.

There are many interesting areas for further research on the topic of modelling non-target effects of control. The dynamics model of poisoning described in section 6.3.1 can include non-target effects but has not been used for that to date. There is a need for better links between the models and field data. There is scope for better linking of models of non-target effects of control with the models of minimum viable populations and effective population size described in section 5.3. Then the ecological significance of non-target effects can be examined in more detail.

6.14 Conclusion

This chapter has given an overview of modelling of vertebrate pest control. It is not intended to be exhaustive but to illustrate the range of analyses that have been used or that could be used. The models can be studied in the same manner as the models described in Chapter 5, and similarly evaluated by subjecting them to the questions listed in Table 5.6. The models in this chapter have an emphasis on control rather than the economics of control. That reflects the bias in the literature so there is a great need for more development and testing of economic models.

The use of models of control, or of damage, requires identifying the key processes occurring and then adapting a general process to the particular process under study; for example, choosing the appropriate general control function listed in Table 6.1 when modelling hunting of particular pests, or using field data to show which control function best explains the data. This says nothing more than if you wish to drive a vehicle to a field site, check before you leave your laboratory whether you need a conventional vehicle or a four-wheel drive. Similarly, if you get stuck in the mud, ask whether it was your vehicle, the road or your driving that got you there. Vehicles, like models, are designed for specific uses and users.

The modelling of control methods has usually concentrated on one control method at a time. The effects of using more than one method sequentially, or simultaneously, should be studied as reported by Carpenter (1981) who studied biological control using pathogens and pesticides simultaneously.

The review of analysis of vertebrate pest control concludes in the next chapter with some brief comments on the overall topic.

7

Conclusion

The aim of this book, as described in Chapter 1, was to review critically the literature on the range of analyses used in vertebrate pest research. I will now briefly summarise the results.

There is a great range of analyses that have been used. The analyses have been used to estimate the level and variation of pest damage and response to control. The three sets of analyses described in this book – statistical, economic and mathematical modelling – have received differing use in research on vertebrate pests. In order of decreasing use the analyses appear to be statistical, modelling and economic. That is surprising as the fundamental reason for the pest status of most animals is their economic impact. There is a great need for closer liaison between biologists and economists to overcome this apparent difference. This could be achieved by involving an economist in a study at the same times as involving a biometrician – project design, analysis and writing-up. The statistical basis of analysis of vertebrate pest control may appear to be emphasised in this book. That should not be interpreted as a signal of my preferences. I believe all three analyses should receive more study.

There have been few detailed comparisons of particular methods within each of the three sets of analyses. For example, predation can be estimated by one or more methods, as described in Chapter 2. However, the biases and cost-effectiveness of methods for estimating the response of predation to control of predators has received less study. The same statement can be made for other topics such as infectious diseases, rodents and bird pests. Key criteria for distinguishing between alternative methods are sensitivity, cost-effectiveness and knowledge of the methods. Just as there are different measures of pest damage (Headley, 1972b), there can be different measures of responses to control. There may be no single answer to the question, 'what is the best analysis to use?' What is clearer is the need to consider the type of analysis used and the need to distinguish between the response of damage and the response of populations to control. A major conclusion of this book is the need for more comparative research on methods for estimating damage and response to control, and for research on comparative analysis of data to complement that already done (Table 7.1). It was beyond the scope of this book to do such comparative research but it needs to be done.

Each analysis has strengths and limitations. At the risk of sounding too negative I will focus on some of the limitations. Statistical analyses have an aura proportional to the number of asterisks cited. However, in the analysis of vertebrate pest control four features stand out as needing closer attention. Firstly, the conclusions from a test are limited by the sample sizes used, and often the sample sizes are very small. Secondly, there is often no testing of alternative hypotheses, for example an hypothesis predicting a straight line compared with an hypothesis predicting a curve. I may appear to have a passion for curves compared with straight lines. Not so. I do, however, remember the damage done in ecology by the original Lotka–Volterra model, which had linear functional and numerical responses. Such linearity is now recognised as simplistic and has been replaced by curvilinear relationships (May, 1981b).

A third limitation of statistical analysis is that tests of linearity can be difficult

Table 7.1. *Some studies which have reported comparative research into methods and analyses relevant to vertebrate pest control*

Topic	Source
Methods for estimating:	
Predation	Schaefer *et al.* (1981)
Bird damage to crops	Dolbeer (1975)
Rate of increase	Eberhardt (1987); Eberhardt & Simmons (1992)
Parameters in random predator equation (functional response)	Juliano & Williams (1987)
Effective population size	Harris & Allendorf (1989)
Analysis of response to control of manipulative experiments	Huson (1980); Skalski & Robson (1992)
Cost–benefit analysis	Sinden (1980)
Methods for studying population dynamics	Krebs (1988)
Models of population growth	Barlow & Clout (1983); Crawley (1983); Eberhardt (1988)

to interpret if the data include a limited range of variation. Tests of linearity would be aided by better use of experimental designs as discussed in Chapter 2. Finally, too many studies have reported data, done no statistical analysis and made important, but unjustified, management conclusions. My concern is that unanalysed results can evolve to become 'conventional wisdom'.

The application of economic analyses and mathematical modelling to vertebrate pest research is encouraging but restricted. Some detailed economic analyses have been reported, for example by Collins *et al.* (1984). Many biologists have only superficial knowledge of either economics, mathematics or modelling. Hence, some immediately neglect and despise the analyses, or they cannot understand them even when they want to. I hope this book has gone some way towards reducing the anger and the anguish.

Emphasis has been on two topics within each of the three sets of analyses described in this book. The first emphasis has been to describe and investigate relationships rather than be concerned with analysis of differences. Hence, control and analysis of control should be an issue of 'how much' rather than a simplistic 'yes or no' issue. Analyses should be concerned more with tests of relationships, such as regression, than tests of difference, for example Student's *t* test. The limited use in vertebrate pest control of response curve studies is not because the idea is new. Preece (1990) cites a response curve study in agronomy from 1793. Problems with logistics are probably a more common reason. Economic analyses depict a variety of relationships between costs and benefits but those relationships have scarcely been applied to empirical data for vertebrate pests. This may be because few studies have estimated the relationship between pest abundance and damage and hence estimated the reduction in damage (benefits) associated with control (costs).

A second emphasis was the blending of theory and practice in the planning and evaluation of vertebrate pest control. The existing empirical emphasis can be enhanced by application of ecological and epidemiological theory. Vertebrate pest control should increasingly be a testing ground for such theory, especially foraging theory, theory of the spread of infectious diseases and predator–prey theory. I support the views of Macfadyen (1975) and Stenseth (1984) for closer links between theory and field data in ecology. The application of cost–benefit analysis and decision theory to the economic evaluation of control is encouraged. These approaches can be at the single species level or the community level with the need in the latter case to estimate costs of effects on non-target species.

At the risk of sounding pretentious may I suggest that there are several fundamental relationships in vertebrate pest control that need further research. First, the relationship between pest abundance and pest damage. Hence, second, the response of pest damage to changes in pest abundance after pest control. Third, the spatial frequency distribution of pest damage. Fourth, the response to pest control of the spatial frequency distribution of damage. Fifth, the relationship between the level of control effort and number of pests removed. Sixth, the level of costs of pest control that maximise the economic benefits of control.

Theoretical models can be linked to pest control practice. Scientists analysing aspects of vertebrate pest control do not have to collect data in an intellectual vacuum. Theory can predict results or suggest alternative sets of results and so indicate which data should be collected and which hypotheses should be tested. The link between theory and practice is often limited in the literature by poor explanation of models. Authors should clearly present how any theory can be tested by stating what data are needed and what tests need to be used. In Chapters 5 and 6, the figures show relationships that can be used to test the models which need more thorough testing. Becker (1979) reported an extreme example of lack of testing of mathematical models. Of 75 scientific papers on models of disease dynamics and control published since the beginning of 1974, only five had any data. Not all of the models described in Chapters 5 and 6 of this book had data attached, either to estimate parameters or as part of tests against field data. That is unfortunate, but hopefully will soon be remedied.

We could probably benefit by looking at the same topics in a related field, for example in weed research. Vertebrate pests and weeds are obviously different, but aspects of analysis in research are similar. The review of modelling in weed management (Doyle, 1991) discussed many of the same issues explored here in Chapters 5 and 6.

Spatial and temporal sources of variation can be identified and incorporated into planning and analysis of control. They can be linked through a planning framework, highlighting the uses and outcomes of the results. Geographic information systems are an obvious starting point and if linked to process models could be a powerful analytical tool.

Research into the many facets of vertebrate pest control can produce interesting and useful results. However, if those results are not presented to the potential end-users in a satisfactory manner then only science benefits and not the people suffering the damage. Grafton, Walters & Bertelsen (1990) described the need for effective extension of research results to the end-users. If that was not done the background research could be viewed simply as a cost with little economic benefit.

I am encouraged by the breadth of the analyses and often by their application and I am optimistic that the analyses will be more integrated in the future. To return to the words of Wittgenstein (1967) described in the Preface, I hope readers have been stimulated to explore their own ideas on analysis of vertebrate pest control.

REFERENCES

Abrams, P. A. (1982). Functional responses of optimal foragers. *American Naturalist*, **120**, 382–90.

Advani, R. & Mathur, R. P. (1982). Experimental reduction of rodent damage to vegetable crops in Indian villages. *Agro-Ecosystems*, **8**, 39–45.

Amlaner, C. J. & Macdonald, D. W. (1980). *A Handbook of Biotelemetry and Radio-tracking*. Oxford: Pergamon Press.

Anderson, R. M. (1981). Population ecology of infectious disease agents. In *Theoretical Ecology: Principles and Applications*, 2nd edn, ed. R. M. May, pp. 318–55. London: Blackwell Scientific Publications.

Anderson, R. M. (1982a). Theoretical basis for the use of pathogens as biological control agents of pest species. *Parasitology*, **84**, 3–33.

Anderson, R. M. (1982b). Directly transmitted viral and bacterial infections of man. In *The Population Dynamics of Infectious Diseases: Theory and Applications*, ed. R. M. Anderson, pp. 1–37. London: Chapman & Hall.

Anderson, R. M. (1982c). Fox rabies. In *The Population Dynamics of Infectious Diseases: Theory and Applications*, ed. R. M. Anderson, pp. 242–61. London: Chapman & Hall.

Anderson, R. M. (1982d). The population dynamics and control of hookworm and roundworm infections. In *The Population Dynamics of Infectious Diseases: Theory and Applications*, ed. R. M. Anderson, pp. 67–108. London: Chapman & Hall.

Anderson, R. M. (1984). Strategies for the control of infectious disease agents. In *Pest and Pathogen Control: Strategic, Tactical and Policy Models*, ed. G. R. Conway, pp. 109–41. Chichester: John Wiley & Sons.

Anderson, R. M. (1986). Vaccination of wildlife reservoirs. *Nature*, **322**, 304–5.

Anderson, R. M., Jackson, H. C., May, R. M. & Smith, A. M. (1981). Population dynamics of fox rabies in Europe. *Nature*, **289**, 765–71.

Anderson, R. M. & May, R. M. (1979). Population biology of infectious diseases: Part I. *Nature*, **280**, 361–7.

Anderson, R. M. & May, R. M. (1982a). Coevolution of hosts and parasites. *Parasitology*, **85**, 411–26.

Anderson, R. M. & May, R. M. (1982b). Directly transmitted infectious diseases: control by vaccination. *Science*, **215**, 1053–60.

Anderson, R. M. & May, R. M. (1985). Vaccination and herd immunity to infectious diseases. *Nature*, **318**, 323–9.

Anderson, R. M. & May, R. M. (1986). The invasion, persistence and spread of infectious diseases within animal and plant communities. *Philosophical Transactions of the Royal Society London B*, **314**, 533–70.

Anderson, R. M. & May, R. M. (1991). *Infectious Diseases of Humans: Dynamics and Control*. Oxford: Oxford University Press.

Anderson, R. M. & Trewhella, W. (1985). Population dynamics of the badger (*Meles meles*) and the epidemiology of bovine tuberculosis (*Mycobacterium bovis*). *Philosophical Transactions of the Royal Society London B*, **310**, 327–81.

Andrzejewski, R. & Jezierski, W. (1978). Management of a wild boar population and its effects on commercial land. *Acta Theriologica*, **23**, 309–39.

Arnold, G. W., Steven, D. E. & Weeldenburg, J. R. (1989). The use of surrounding farmland by western grey kangaroos living in a remnant of wandoo woodland and their impact on crop production. *Australian Wildlife Research*, **16**, 85–93.

Arthur, C. P. & Louzis, C. (1988). A review of myxomatosis among rabbits in France. *Reviews of Science and Technology Office International Epizootic*, **7**, 959–76.

Auld, B. A. & Tisdell, C. A. (1986). Impact assessment of biological invasions. In *Ecology of Biological Invasions: An Australian Perspective*, ed. R. H. Groves & J. J. Burdon, pp. 79–88. Canberra: Australian Academy Science.

Bacon, P. J. (1981). The consequences on unreported fox rabies. *Journal of Environmental Management*, **13**, 195–200.

Bacon, P. J. (1985). *Population Dynamics of Rabies in Wildlife*. London: Academic Press.

Bacon, P. J. & Macdonald, D. W. (1980). To control rabies: vaccinate foxes. *New Scientist*, **87** (1216), 640–5.

Bailey, N. T. J. (1975). *The Mathematical Theory of Infectious Diseases and its Applications*. 2nd edn. New York: Hafner Press.

Baker, S. J. & Clarke, C. N. (1988). Cage trapping coypus (*Myocastor coypus*) on baited rafts. *Journal of Applied Ecology*, **25**, 41–8.

Balasubramanyam, M., Christopher, M. J. & Purushotham, K. R. (1985). Field evaluation of three anticoagulant rodenticides against rodent pests in paddy fields. *Tropical Pest Management*, **31**, 299–301.

Ball, F. G. (1985). Spatial models for the spread and control of rabies incorporating group size. In *Population Dynamics of Rabies*, ed. P. J. Bacon, pp. 197–222. London: Academic Press.

Bamford, J. (1970). Evaluating opossum poisoning operations by interference with non-toxic baits. *Proceedings of the New Zealand Ecological Society*, **17**, 118–25.

Barlow, N. D. (1991a). A spatially aggregated disease/host model for bovine Tb in New Zealand possum populations. *Journal of Applied Ecology*, **28**, 777–93.

Barlow, N. D. (1991b). Control of endemic bovine Tb in New Zealand possum populations: results from a simple model. *Journal of Applied Ecology*, **28**, 794–809.

Barlow, N. D. (1993). A model for the spread of bovine Tb in New Zealand possum populations. *Journal of Applied Ecology*, **30**, 156–64.

Barlow, N. D. & Clout, M. N. (1983). A comparison of 3-parameter, single-species population models, in relation to the management of brushtail possums in New Zealand. *Oecologia*, **60**, 250–8.

Barnett, S. A. (1958). Experiments on 'neophobic' response in wild and laboratory rats. *British Journal of Psychology*, **49**, 195–201.

Barnett, S. A. & Prakash, I. (1976). *Rodents of Economic Importance*. London: Heinemann.

Basson, M., Beddington, J. R. & May, R. M. (1991). An assessment of the maximum sustained yield of ivory from African elephant populations. *Mathematical Biosciences*, **104**, 73–95.

Batcheler, C. L. (1982). Quantifying 'bait quality' from number of random encounters required to kill a pest. *New Zealand Journal of Ecology*, **5**, 129–39.

Batcheler, C. L., Darwin, J. H. & Pracy, L. T. (1967). Estimation of opposum (*Trichosurus vulpecula*) populations and results of poison trials from trapping data. *New Zealand Journal of Science*, **10**, 97–114.

Bayliss, P. (1985). The population dynamics of red and western grey kangaroos in arid New South Wales, Australia. II. The numerical response function. *Journal of Animal Ecology*, **54**, 127–35.

Bayliss, P. (1987). Kangaroo dynamics. In *Kangaroos: Their Ecology and Management in the Sheep Rangelands of Australia*, ed. G. Caughley, N. Shepherd & J. Short, pp. 119–34. Cambridge: Cambridge University Press.

Bayliss, P. & Yeomans, K. M. (1989). Distribution and abundance of feral livestock in the 'Top End' of the Northern Territory (1985–86), and their relation to population control. *Australian Wildlife Research*, **16**, 651–76.

Becker, N. G. (1977). On a general stochastic epidemic model. *Theoretical Population Biology*, **11**, 23–36.

Becker, N. G. (1979). The uses of epidemic models. *Biometrics*, **35**, 295–305.

Begon, M., Harper, J. L. & Townsend, C. R. (1990). *Ecology: Individuals, Populations and Communities*, 2nd edn. Oxford: Blackwell Scientific Publications.

Belden, R. C. & Pelton, M. R. (1975). European wild hog rooting in the mountains of east Tennessee. In *Proceedings of the Twenty-Ninth Annual Conference of Southeast Association Game and Fish Commissioners*, pp. 665–71. Saint Louis.

Bell, R. (1983). Deciding what to do and how to do it. In *Guidelines for the Management of Large Mammals in African Conservation Areas*, ed. A. A. Ferrar, pp. 51–75. Pretoria: South African National Scientific Programmes. Report No. 69.

Belsky, A. J. (1986). Does herbivory benefit plants? A review of the evidence. *American Naturalist*, **127**, 870–92.

Berger, J. (1990). Persistence of different-sized populations: an empirical assessment of rapid extinctions in bighorn sheep. *Conservation Biology*, **4**, 91–8.

Bergerud, A. T., Wyett, W. & Snider, B. (1983). The role of wolf predation in limiting a moose population. *Journal of Wildlife Management*, **47**, 977–88.

Bernhardt, G. E. & Seamans, T. W. (1990). Red-winged blackbird feeding behavior on two sweet corn cultivars. *Wildlife Society Bulletin*, **18**, 83–6.

Birley, M. H. (1979). The theoretical control of seasonal pests: a single species model. *Mathematical Biosciences*, **43**, 141–57.

Bjorge, R. R. & Gunson, J. R. (1985). Evaluation of wolf control to reduce cattle predation in Alberta. *Journal of Range Management*, **38**, 483–7.

Blancou, J., Kieny, M. P., Lathe, R., Lecocq, J. P., Pastoret, P. P., Soulebot, J. P. & Desmettre, P. (1986). Oral vaccination of the fox against rabies using a live recombinant vaccinia virus. *Nature*, **322**, 373–5.

Blancou, J., Pastoret, P. P., Brochier, B., Thomas, I. & Bogel, K. (1988). Vaccinating wild animals against rabies. *Reviews of Science and Technology Office International Epizootic*, **7**, 1005–13.

Boggess, E. K., Andrews, R. D. & Bishop, R. A. (1978). Domestic animal losses to coyotes and dogs in Iowa. *Journal of Wildlife Management*, **42**, 362–72.

Boiko, A. I., Kruglikov, B. A. & Shchenev, A. I. (1987). Organisation of measures against foot-and-mouth disease in the virus migration zone. *Veterinariya*, **10**, 33–4.

Bomford, M. (1990a). Ineffectiveness of a sonic device for deterring starlings. *Wildlife Society Bulletin*, **18**, 151–6.

Bomford, M. (1990b). *A Role for Fertility Control in Wildlife Management*. Canberra: Bureau of Rural Resources.

Bomford, M. & O'Brien, P.H. (1990). Sonic deterrents in animal damage control: a review of device tests and effectiveness. *Wildlife Society Bulletin*, **18**, 411–22.

Boulton, W.J. & Freeland, W.J. (1991). Models for the control of feral water buffalo (*Bubalus bubalis*) using constant levels of offtake and effort. *Wildlife Research*, **18**, 63–73.

Bourne, J. & Dorrance, M.J. (1982). A field test of lithium chloride aversion to reduce coyote predation on domestic sheep. *Journal of Wildlife Management*, **46**, 235–9.

Braithwaite, L.W., Turner, J. & Kelly, J. (1984). Studies on the arboreal marsupial fauna of eucalypt forests being harvested for woodpulp at Eden, N.S.W. III. Relationships between faunal densities, eucalypt occurrence and foliage nutrients, and soil parent materials. *Australian Wildlife Research*, **11**, 41–8.

Bradley, D.J. (1982). Epidemiological models: theory and reality. In *The Population Dynamics of Infectious Diseases: Theory and Applications*, ed. R.M. Anderson, pp. 320–33. London: Chapman & Hall.

Bratton, S.P. (1974). The effect of the European wild boar (*Sus scrofa*) on the high-elevation vernal flora in Great Smoky Mountains National Park. *Bulletin of the Torrey Botanical Club*, **101**, 198–206.

Breckwoldt, R. (1983). *Wildlife in the Home Paddock*. Sydney: Angus & Robertson.

Briggs, S.V. & Holmes, J.E. (1988). Bag sizes of waterfowl in New South Wales and their relation to antecedent rainfall. *Australian Wildlife Research*, **15**, 459–68.

Brochier, B.M., Languet, B., Blancou, J., Kieny, M.P., Lecocq, J.P., Costy, F., Desmettre, P. & Pastoret, P. (1988a). Use of recombinant vaccinia-rabies virus for oral vaccination of fox cubs (*Vulpes vulpes*, L) against rabies. *Veterinary Microbiology*, **18**, 103–8.

Brochier, B., Thomas, I., Iokem, A., Ginter, A., Kalpers, J., Paquot, A., Costy, F. & Pastoret, P. (1988b). A field trial in Belgium to control fox rabies by oral immunisation. *Veterinary Record*, **123**, 618–21.

Brockie, R.E., Loope, L.L., Usher, M.B. & Hamann, O. (1988). Biological invasions of island nature reserves. *Biological Conservation*, **44**, 9–36.

Brothers, N.P., Eberhard, I.E., Copson, G.R. & Skira, I.J. (1982). Control of rabbits on Macquarie Island by myxomatosis. *Australian Wildlife Research*, **9**, 477–85.

Brough, T. & Bridgman, C.J. (1980). An evaluation of long grass as a bird deterrent on British airfields. *Journal of Applied Ecology*, **17**, 243–53.

Bruggers, R.L. (1989). Assessment of bird-repellent chemicals in Africa. In *Quelea Quelea: Africa's Bird Pest*, ed. R.L. Bruggers & C.C.H. Elliott, pp. 262–80. Oxford: Oxford University Press.

Bruggers, R.L. & Elliott, C.C.H. (1989). *Quelea Quelea: Africa's Bird Pest*. Oxford: Oxford University Press.

Brunner, H. & Coman, B.J. (1983). The ingestion of artificially coloured grain by birds, and its relevance to vertebrate pest control. *Australian Wildlife Research*, **10**, 303–10.

Bryant, H., Hone, J. & Nicholls, P. (1984). The acceptance of dyed grain by feral pigs and birds. I. Birds. *Australian Wildlife Research*, **11**, 509–16.

Buckland, S.T., Anderson, D.R., Burnham, K.P. & Laake, J.L. (1993). *Distance Sampling: Estimating Abundance of Biological Populations*. London: Chapman & Hall.

Buckle, A.P. (1985). Field trials of a new sub-acute rodenticide flupropadine, against wild Norway rats (*Rattus norvegicus*). *Journal of Hygiene*, **95**, 505–12.

Buckle, A.P., Rowe, F.P. & Husin, A.R. (1984). Field trials of warfarin and brodifacoum wax block baits for the control of the rice field rat, *Rattus argentiventer*, in peninsular Malaysia. *Tropical Pest Management*, **30**, 51–8.

Buechner, M. (1987). Conservation in insular parks: simulation models of factors affecting the movement of animals across park boundaries. *Biological Conservation*, **41**, 57–76.

Burger, J. (1985). Factors affecting bird strikes on aircraft at a coastal airport. *Biological*

Conservation, **33**, 1–28.
Burgman, M. A., Akcakaya, H. R. & Loew, S. S. (1988). The use of extinction models for species conservation. *Biological Conservation*, **43**, 9–25.
Burns, R. J. (1983). Microencapsulated lithium chloride bait aversion did not stop coyote predation on sheep. *Journal of Wildlife Management*, **47**, 1010–7.
Caley, P. (1993). The ecology and management of feral pigs in the 'wet–dry' tropics of the Northern Territory. M. App. Sc. thesis. Canberra: University of Canberra.
Cannon, R. M. & Roe, R. T. (1982). *Livestock Disease Surveys: A Field Manual for Veterinarians*. Canberra: Australian Government Publishing Service.
Carpenter, S. R. (1981). Effect of control measures on pest populations subject to regulation by parasites and pathogens. *Journal of Theoretical Biology*, **92**, 181–4.
Caslick, J. W. & Decker, D. J. (1979). Economic feasibility of a deer-proof fence for apple orchards. *Wildlife Society Bulletin*, **7**, 173–5.
Caughley, G. (1970a). Liberation, dispersal and distribution of Himalayan thar (*Hemitragus jemlahicus*) in New Zealand. *New Zealand Journal of Science*, **13**, 220–39.
Caughley, G. (1970b). Eruption of ungulate populations, with emphasis on Himalayan thar in New Zealand. *Ecology*, **51**, 53–72.
Caughley, G. (1976). Wildlife management and the dynamics of ungulate populations. In *Applied Biology*, Vol. 1, ed. T. H. Coaker, pp. 183–246. London: Academic Press.
Caughley, G. (1977). Sampling in aerial survey. *Journal of Wildlife Management*, **41**, 605–15.
Caughley, G. (1980). *Analysis of Vertebrate Populations*. Reprinted with corrections. London: John Wiley & Sons.
Caughley, G. (1981). Overpopulation. In *Problems in Management of Locally Abundant Wild Animals*, ed. P. A. Jewell & S. Holt, pp. 7–19. New York: Academic Press.
Caughley, G. (1987). Ecological relationships. In *Kangaroos: Their Ecology and Management in the Sheep Rangelands of Australia*, ed. G. Caughley, N. Shepherd & J. Short, pp. 159–87. Cambridge: Cambridge University Press.
Caughley, G., Dublin, H. & Parker, I. (1990). Projected decline of the African elephant. *Biological Conservation*, **54**, 157–64.
Caughley, G. & Krebs, C. J. (1983). Are big mammals simply little mammals writ large? *Oecologia*, **59**, 7–17.
Caughley, G. & Lawton, J. H. (1981). Plant–herbivore systems. In *Theoretical Ecology: Principles and Applications*, 2nd edn, ed. R. M. May, pp. 132–66. London: Blackwell Scientific Publications.
Caughley, G., Pech, R. & Grice, D. (1992). Effect of fertility control on a population's productivity. *Wildlife Research*, **19**, 623–7.
Challies, C. N. (1975). Feral pigs (*Sus scrofa*) on Auckland Island: status, and effects on vegetation and nesting sea birds. *New Zealand Journal of Zoology*, **2**, 479–90.
Chapman, C. A., Chapman, L. J. & McLaughlin, R. L. (1989). Multiple central place foraging by spider monkeys: travel consequences of using many sleeping sites. *Oecologia*, **79**, 506–11.
Charnov, E. L. (1976). Optimal foraging: the marginal value theorem. *Theoretical Population Biology*, **9**, 129–36.
Cheeseman, C. L., Jones, G. W., Gallagher, J. & Mallinson, P. J. (1981). The population structure, density and prevalence of tuberculosis (*Mycobacterium bovis*) in badgers (*Meles meles*) from four areas in south-west England. *Journal of Applied Ecology*, **18**, 795–804.
Cheeseman, C. L., Wilesmith, J. W. & Stuart, F. A. (1989). Tuberculosis: the disease and its epidemiology in the badger, a review. *Epidemiology and Infection*, **103**, 113–25.
Cheeseman, C. L., Wilesmith, J. W., Stuart, F. A. & Mallinson, P. J. (1988). Dynamics of tuberculosis in a naturally infected badger population. *Mammal Review*, **18**, 61–72.

Cherrett, J. M., Ford, J. B., Herbert, I. V. & Probert, A. J. (1971). *The Control of Injurious Animals*. London: English Universities Press.
Chiang, H. C. (1979). A general model of the economic threshold level of pest populations. *Food and Agriculture Organisation Plant Protection Bulletin*, **27**, 71–3.
Choquenot, D. (1988). Feral donkeys in northern Australia: population dynamics and the cost of control. M. App. Sc. thesis. Canberra: Canberra College of Advanced Education.
Choquenot, D. (1990). Rate of increase for populations of feral donkeys in northern Australia. *Journal of Mammalogy*, **71**, 151–5.
Choquenot, D. (1991). Density-dependent growth, body condition, and demography in feral donkeys: testing the food hypothesis. *Ecology*, **72**, 805–13.
Choquenot, D., Kay, B. & Lukins, B. (1990). An evaluation of warfarin for the control of feral pigs. *Journal of Wildlife Management*, **54**, 353–9.
Choquenot, D., Kilgour, R. J. & Lukins, B. S. (1993). An evaluation of feral pig trapping. *Wildlife Research*, **20**, 15–22.
Clark, C. W. (1976). *Mathematical Bioeconomics: The Optimal Management of Renewable Resources*. New York: John Wiley & Sons.
Clark, C. W. (1981). Bioeconomics. In *Theoretical Ecology: Principles and Applications*, 2nd edn, ed. R. M. May, pp. 387–418. Oxford: Blackwell Scientific Publications.
Clout, M. N. & Barlow, N. D. (1982). Exploitation of brushtail possum populations in theory and practice. *New Zealand Journal of Ecology*, **5**, 29–35.
Cochran, W. G. (1977). *Sampling Techniques*, 3rd edn. New York: John Wiley & Sons.
Cochran, W. G. & Cox, G. M. (1957). *Experimental Designs*, 2nd edn. New York: John Wiley & Sons.
Cohen, J. E. (1973). Selective host mortality in a catalytic model applied to schistosomiasis. *American Naturalist*, **107**, 199–212.
Cohen, J. E. (1976). Schistosomiasis: a human host–parasite system. In *Theoretical Ecology: Principles and Applications*, 1st edn, ed. R. M. May, pp. 237–56. Oxford: Blackwell Scientific Publications.
Coleman, J. D. (1988). Distribution, prevalence, and epidemiology of bovine tuberculosis in brushtail possums, *Trichosurus vulpecula*, in the Hohonu range, New Zealand. *Australian Wildlife Research*, **15**, 651–63.
Collins, A. R., Workman, J. P. & Uresk, D. W. (1984). An economic analysis of black-tailed prairie dog (*Cynomys ludovicianus*) control. *Journal of Range Management*, **37**, 358–61.
Conley, R. H. (1977). Management and research of the European wild hog in Tennessee. In *Research and Management of Wild Hog Populations*, ed. G. W. Wood, pp. 67–70. Georgetown: Belle Baruch Forest Science Institute.
Connell, J. H. (1978). Diversity in tropical rain forests and coral reefs. *Science*, **199**, 1302–10.
Conover, M. R. (1990). Reducing mammalian predation on eggs by using a conditioned taste aversion to deceive predators. *Journal of Wildlife Management*, **54**, 360–5.
Conover, M. R., Francik, J. G. & Miller, D. E. (1977). An experimental evaluation of aversive conditioning for controlling coyote predation. *Journal of Wildlife Management*, **41**, 775–9.
Conover, M. R., Francik, J. G. & Miller, D. E. (1979). Aversive conditioning in coyotes: a reply. *Journal of Wildlife Management*, **43**, 209–11.
Conway, G. R. (1977). Mathematical models in applied ecology. *Nature*, **269**, 291–7.
Conway, G. (1981). Man versus pests. In *Theoretical Ecology: Principles and Applications*, 2nd edn, ed. R. M. May, pp. 356–86. Oxford: Blackwell Scientific Publications.
Conway, G. (1984). Introduction. In *Pest and Pathogen Control: Strategic, Tactical and Policy Models*, ed. G. R. Conway, pp. 1–11. New York: John Wiley & Sons.

Conway, G. R. & Comins, H. N. (1979). Resistance to pesticides. 2. Lessons in strategy from mathematical models. *Span*, **22**, 53–5.

Cooke, B. D. (1981). Rabbit control and the conservation of native mallee vegetation on roadsides in South Australia. *Australian Wildlife Research*, **8**, 627–36.

Cooke, B. D. (1983). Changes in the age-structure and size of populations of wild rabbits in South Australia, following the introduction of European rabbit fleas, *Spilopsyllus cuniculi* (Dale), as vectors of myxomatosis. *Australian Wildlife Research*, **10**, 105–20.

Cooke, B. D. (1987). The effects of rabbit grazing on regeneration of sheoaks, *Allocasuarina verticilliata* and saltwater ti-trees, *Melaleuca halmaturorum*, in the Coorong National Park, South Australia. *Australian Journal of Ecology*, **13**, 11–20.

Cooke, B. D. & Hunt, L. P. (1987). Practical and economic aspects of rabbit control in hilly semiarid South Australia. *Australian Wildlife Research*, **14**, 219–23.

Cooray, R. G. & Mueller–Dombois, D. (1981). Feral pig activity. In *Island Ecosystems: Biological Organisation in Selected Hawaiian Communities*, ed. D. Mueller-Dombois, K. W. Bridges & H. L. Carson, pp. 309–19. Stroudsburg: Hutchinson Research Publishing Company.

Coulson, R. N. (1992). Intelligent geographic information systems and integrated pest management. *Crop Protection*, **11**, 507–16.

Cowan, D. P., Vaughan, J. A. & Christer, W. G. (1987). Bait consumption by the European rabbit in southern England. *Journal of Wildlife Management*, **51**, 386–92.

Cowan, D. P., Vaughan, J. A., Prout, K. J. & Christer, W. G. (1984). Markers for measuring bait consumption by the European wild rabbit. *Journal of Wildlife Management*, **48**, 1403–9.

Cowie, R. J. (1977). Optimal foraging in great tits *Parus major*. *Nature*, **268**, 137–9.

Coyne, M. J., Smith, G. & McAllister, F. E. (1989). Mathematic model for the population biology of rabies in raccoons in the mid-Atlantic states. *American Journal of Veterinary Research*, **50**, 2148–54.

Crabb, A. C., Salmon, T. P. & Marsh, R. E. (1986). Bird problems in California pistachio production. In *Proceedings of the Twelfth Vertebrate Pest Conference*, ed. T. P. Salmon, pp. 295–302. Davis: University of California.

Crawley, M. J. (1983). *Herbivory: The Dynamics of Animal–Plant Interactions*. London: Blackwell Scientific Publications.

Crawley, M. J. (1986). The population biology of invaders. *Philosophical Transactions of the Royal Society London B*, **314**, 711–31.

Crawley, M. J. & Weiner, J. (1991). Plant size variation and vertebrate herbivory: winter wheat grazed by rabbits. *Journal of Applied Ecology*, **28**, 154–72.

Crisp, M. D. & Lange, R. T. (1976). Age-structure, distribution and survival under grazing of the arid-zone shrub *Acacia birkittii*. *Oikos*, **27**, 86–92.

Croft, J. D. (1990). The impact of rabbits on sheep production. M.Sc. thesis. Sydney: University of New South Wales.

Croft, J. D. & Hone, L. J. (1978). The stomach contents of foxes, *Vulpes vulpes*, collected in New South Wales. *Australian Wildlife Research*, **5**, 85–92.

Crosbie, S. F., Laas, F. J., Godfrey, M. E. R., Williams, J. M. & Moore, D. S. (1986). A field assessment of the anticoagulant brodifacoum against rabbits, *Oryctolagus cuniculus*. *Australian Wildlife Research*, **13**, 189–95.

Cummings, J. L., Guarino, J. L. & Knittle, C. E. (1989). Chronology of blackbird damage to sunflowers. *Wildlife Society Bulletin*, **17**, 50–2.

Dahlsten, D. L. (1986). Control of invaders. In *Ecology of Biological Invasions of North America and Hawaii*, ed. H. A. Mooney & J. A. Drake, pp. 275–302. New York: Springer-Verlag.

Dale, V. H., Franklin, R. L. A., Post, W. M. & Gardner, R. H. (1991). Sampling ecological information: choice of sample size. *Ecological Modelling*, **57**, 1–10.

Daly, J. C. (1980). Age, sex and season: factors which determine the trap response of the European wild rabbit, *Oryctolagus cuniculus*. *Australian Wildlife Research*, **7**, 421–32.

Damuth, J. (1981). Population density and body size in mammals. *Nature*, **290**, 699–700.

Davis, D. E., Myers, K. & Hoy, J. B. (1976). Biological control among vertebrates. In *Theory and Practice of Biological Control*, ed. C. B. Huffaker & P. S. Messenger, pp. 501–19. New York: Academic Press.

Dawson, D. G. (1970). Estimation of grain loss due to sparrows (*Passer domesticus*) in New Zealand. *New Zealand Journal of Agricultural Research*, **13**, 681–8.

Dawson, D. G. & Bull, P. C. (1970). A questionnaire survey of bird damage to fruit. *New Zealand Journal of Agricultural Research*, **13**, 362–71.

De Grazio, J. W. (1989). Pest birds: an international perspective. In *Quelea Quelea: Africa's Bird Pest*, ed. R. L. Bruggers & C. C. H. Elliott, pp. 1–8. Oxford: Oxford University Press.

Dillon, J. L. (1977). *The Analysis of Response in Crop and Livestock Production*, 2nd edn. Oxford: Pergamon Press.

Dobson, A. J. (1983). *Introduction to Statistical Modelling*. London: Chapman & Hall.

Dobson, A. P. (1988). Restoring island ecosystems: the potential role of parasites to control introduced mammals. *Conservation Biology*, **2**, 31–9.

Dobson, A. P. & May, R. M. (1986). Patterns of invasions by pathogens and parasites. In *Ecology of Invasions of North America and Hawaii*, ed. H. A. Mooney & J. A. Drake, pp. 58–76. New York: Springer-Verlag.

Dolbeer, R. A. (1975). A comparison of two methods for estimating bird damage to sunflowers. *Journal of Wildlife Management*, **39**, 802–6.

Dolbeer, R. A. (1981). Cost–benefit determination of blackbird damage control for cornfields. *Wildlife Society Bulletin*, **9**, 44–51.

Dolbeer, R. A. (1988). Current status and potential of lethal means of reducing bird damage in agriculture. In *Acta XIX Congressus Internationalis Ornithologici*, ed. H. Ouellet, pp. 474–83. Ottawa: University of Ottawa Press.

Donaldson, A. I., Lee, M. & Shimshony, A. (1988). A possible airborne transmission of foot and mouth disease virus from Jordon to Israel: a simulated computer analysis. *Israel Journal of Veterinary Medicine*, **44**, 92–6.

Dorrance, M. J. & Roy, L. D. (1976). Predation losses of domestic sheep in Alberta. *Journal of Range Management*, **29**, 457–60.

Doyle, C. J. (1991). Mathematical models in weed management. *Crop Protection*, **10**, 432–44.

Drummond, D. C. (1985). Developing and monitoring urban rodent control programmes. *Acta Zoologica Fennica*, **173**, 145–8.

Drummond, D. C. & Rennison, B. D. (1973). The detection of rodent resistance to anticoagulants. *Bulletin of the World Health Organisation*, **48**, 239–42.

Dwyer, G. (1991). The roles of density, stage, and patchiness in the transmission of an insect virus. *Ecology*, **72**, 559–74.

Dwyer, G., Levin, S. A. & Buttel, L. (1990). A simulation model of the population dynamics and evolution of myxomatosis. *Ecological Monographs*, **60**, 423–47.

Dyer, M. I. (1975). The effects of red-winged blackbirds (*Agelaius phoeniceus* L.) on biomass production of corn grains (*Zea mays* L.). *Journal of Applied Ecology*, **12**, 719–26.

Dyer, M. I. & Ward, P. (1977). Management of pest situations. In *Granivorous Birds in Ecosystems*, ed. J. Pinowski & S. C. Kendeigh, pp. 267–300. Cambridge: Cambridge University Press.

Eastland, W. G. & Beasom, S. L. (1986). Effects of ambient temperature on the 1080-LD_{50}

of raccoons. *Wildlife Society Bulletin*, **14**, 234–5.
Ebenhard, T. (1988). Introduced birds and mammals and their ecological effects. *Swedish Wildlife Research*, **13**, 1–107.
Eberhardt, L. L. (1976). Quantitative ecology and impact assessment. *Journal of Environmental Management*, **4**, 27–70.
Eberhardt, L. L. (1978). Appraising variability in population studies. *Journal of Wildlife Management*, **42**, 207–38.
Eberhardt, L. L. (1987). Population projections from simple models. *Journal of Applied Ecology*, **24**, 103–18.
Eberhardt, L. L. (1988). Testing hypotheses about populations. *Journal of Wildlife Management*, **52**, 50–6.
Eberhardt, L. L. & Pitcher, K. W. (1992). A further analysis of the Nelchina caribou and wolf data. *Wildlife Society Bulletin*, **20**, 385–95.
Eberhardt, L. L. & Simmons, M. A. (1992). Assessing rates of increase from trend data. *Journal of Wildlife Management*, **56**, 603–10.
Eberhardt, L. L. & Thomas, J. M. (1991). Designing environmental field studies. *Ecological Monographs*, **61**, 53–73.
Elliott, C. C. H. (1989). The pest status of the quelea. In *Quelea Quelea: Africa's Bird Pest*, ed. R. L. Bruggers & C. C. H. Elliott, pp. 17–34. Oxford: Oxford University Press.
Eltringham, S. K. (1984). *Wildlife Resources and Economic Development*. Chichester: John Wiley & Sons.
Feare, C. J., Greig-Smith, P. W. & Inglis, I. R. (1988). Current status and potential of non-lethal means of reducing bird damage in agriculture. In *Acta XIX Congressus Internationalis Ornithologici*, ed. H. Ouellet, pp. 493–506. Ottawa: University of Ottawa Press.
Fenner, F. (1983). Biological control, as exemplified by smallpox eradication and myxomatosis. *Proceedings of the Royal Society London B*, **218**, 259–85.
Fenner, F. & Ratcliffe, F. N. (1965). *Myxomatosis*. Cambridge: Cambridge University Press.
Fennessy, B. V. (1966). The impact of wildlife species on sheep production in Australia. *Proceedings of the Australian Society of Animal Production*, **6**, 148–56.
Finney, D. J. (1971). *Probit Analysis*, 3rd edn. Cambridge: Cambridge University Press.
Fitzgerald, B. M. & Veitch, C. R. (1985). The cats of Herekopare Island, New Zealand: their history, ecology and affects on birdlife. *New Zealand Journal of Zoology*, **12**, 319–30.
Foran, B. D. (1986). The impact of rabbits and cattle on an arid calcareous shrubby grassland in central Australia. *Vegetatio*, **66**, 49–59.
Foran, B. D., Low, W. A. & Strong, B. W. (1985). The response of rabbit populations and vegetation to rabbit control on a calcareous shrubby grassland in central Australia. *Australian Wildlife Research*, **12**, 237–47.
Forster, J. A. (1975). Electric fencing to protect sandwich terns against foxes. *Biological Conservation*, **7**, 85.
Fowler, C. W. (1988). Population dynamics as related to rate of increase per generation. *Evolutionary Ecology*, **2**, 197–204.
Fox, J. R. & Pelton, M. R. (1977). An evaluation of control techniques for the European wild hog in the Great Smoky Mountains National Park. In *Research and Management of Wild Hog Populations*, ed. G. W. Wood, pp. 53–66. Georgetown: Belle Baruch Forest Science Institute.
Francis, C. F. & Thornes, J. B. (1990). Runoff hydrographs from three Mediterranean vegetation cover types. In *Vegetation and Erosion*, ed. J. B. Thornes, pp. 363–84. London: John Wiley & Sons.

Freeland, W. J. (1990). Large herbivorous mammals: exotic species in northern Australia. *Journal of Biogeography*, **17**, 445–9.
Freeland, W. J. & Boulton, W. J. (1990). Feral water buffalo (*Bubalus bubalis*) in the major floodplains of the 'Top End', Northern Territory, Australia: population growth and the brucellosis and tuberculosis eradication campaign. *Australian Wildlife Research*, **17**, 411–20.
Freeland, W. J. & Choquenot, D. (1990). Determinants of herbivore carrying capacity: plants, nutrients, and *Equus asinus* in northern Australia. *Ecology*, **71**, 589–97.
Friend, J. A. (1990). The numbat *Myrmecobius fasciatus* (Myrmecobiidae): history of decline and potential for recovery. *Proceedings of the Ecological Society of Australia*, **16**, 369–77.
Fryxell, J. M., Hussell, D. J. T., Lambert, A. B. & Smith, P. C. (1991). Time lags and population fluctuations in white-tailed deer. *Journal of Wildlife Management*, **55**, 377–85.
Fujii, K., Holling, C. S. & Mace, P. M. (1986). A simple generalised model of attack by predators and parasites. *Ecological Research*, **1**, 141–56.
Garrott, R. A. (1991). Feral horse fertility control: potential and limitations. *Wildlife Society Bulletin*, **19**, 52–8.
Gerrodette, T. (1987). A power analysis for detecting trends. *Ecology*, **68**, 1364–72.
Giles, J. R. (1980). Ecology of feral pigs in New South Wales. Ph.D. thesis. Sydney: University of Sydney.
Gillespie, G. D. (1982). Greenfinch feeding behaviour and impact on a rapeseed crop in Oamaru, New Zealand. *New Zealand Journal of Zoology*, **9**, 481–6.
Gillespie, G. D. (1985). Feeding behaviour and impact of ducks on ripening barley crops grown in Otago, New Zealand. *Journal of Applied Ecology*, **22**, 347–56.
Glahn, J. F. & Otis, D. L. (1986). Factors influencing blackbird and European starling damage at livestock feeding operations. *Journal of Wildlife Management*, **50**, 15–19.
Gloster, J., Blackall, R. M., Sellers, R. F. & Donaldson, A. I. (1981). Forecasting the airborne spread of foot-and-mouth disease. *Veterinary Record*, **108**, 370–4.
Gluesing, E. A., Balph, D. F. & Knowlton, F. F. (1980). Behavioral patterns of domestic sheep and their relationship to coyote predation. *Applied Animal Ethology*, **6**, 315–30.
Goodloe, R. B., Warren, R. J., Cothran, E. G., Bratton, S. P. & Trembicki, K. A. (1991). Genetic variation and its management applications in eastern U.S. feral horses. *Journal of Wildlife Management*, **55**, 412–21.
Gorynska, W. (1981). Method of determining relations between the extent of damage in farm crops, big game numbers, and environmental conditions. *Acta Theriologica*, **26**, 469–81.
Gosling, M. (1989). Extinction to order. *New Scientist*, **121** (1654), 44–9.
Gosling, L. M. & Baker, S. J. (1987). Planning and monitoring an attempt to eradicate coypus from Britain. *Symposium of the Zoological Society of London*, **58**, 99–113.
Gosling, L. M. & Baker, S. J. (1989). The eradication of muskrats and coypus from Britain. *Biological Journal of the Linnean Society*, **38**, 39–51.
Gosling, L. M., Baker, S. J. & Clarke, C. N. (1988). An attempt to remove coypus (*Myocastor coypus*) from a wetland habitat in East Anglia. *Journal of Applied Ecology*, **25**, 49–62.
Gosling, L. M., Baker, S. J. & Skinner, J. R. (1983). A simulation approach to investigating the response of a coypu population to climatic variation. *European Plant Protection Organisation Bulletin*, **13**, 183–92.
Gotceitas, V. & Colgan, P. (1989). Predator foraging success and habitat complexity: quantitative test of the threshold hypothesis. *Oecologia*, **80**, 158–66.
Grafton, R. Q., Walters, E. B. & Bertelsen, M. K. (1990). An evaluation of a FSR/E project: costs and benefits of research and extension. *Agricultural Systems*, **34**, 207–21.

Granett, P., Trout, J.R., Messersmith, D.H. & Stockdale, T.M. (1974). Sampling corn for bird damage. *Journal of Wildlife Management*, **38**, 903–9.

Grant, W.E., Fraser, S.O. & Isakson, K.G. (1984). Effects of vertebrate pesticides on non-target wildlife populations: evaluation through modelling. *Ecological Modelling*, **21**, 85–108.

Greaves, J.H. (1985). The present status of resistance to anticoagulants. *Acta Zoologica Fennica*, **173**, 159–62.

Green, J.S. & Woodruff, R.A. (1983). The use of three breeds of dog to protect rangeland sheep from predators. *Applied Animal Ethology*, **11**, 141–61.

Green, J.S., Woodruff, R.A. & Tueller, T.T. (1984). Livestock-guarding dogs for predator control: costs, benefits and practicality. *Wildlife Society Bulletin*, **12**, 44–50.

Green, R.F. (1984). Stopping rules for optimal foragers. *American Naturalist*, **123**, 30–40.

Green, W.Q. & Coleman, J.D. (1984). Response of a brush-tailed possum population to intensive trapping. *New Zealand Journal of Zoology*, **11**, 319–28.

Greenhalgh, D. (1986). Control of an epidemic spreading in a heterogeneously mixing population. *Mathematical Biosciences*, **80**, 23–45.

Griffith, B., Scott, J.M., Carpenter, J.W. & Reed, C. (1989). Translocation as a species conservation tool: status and strategy. *Science*, **245**, 477–80.

Grossman, S.I. & Turner, J.E. (1974). *Mathematics for the Biological Sciences*. New York: Macmillan.

Gustavson, C.R. (1979). An experimental evaluation of aversive conditioning for controlling coyote predation: a critique. *Journal of Wildlife Management*, **43**, 208–9.

Gustavson, C.R., Garcia, J., Hankins, W.G. & Rusiniak, K.W. (1974). Coyote predation control by aversive conditioning. *Science*, **184**, 581–3.

Gustavson, C.R., Kelly, D.J., Sweeney, M. & Garcia, J. (1976). Prey-lithium aversions. I: Coyotes and wolves. *Behavioral Biology*, **17**, 61–72.

Hairston, N.G. (1989). *Ecological Experiments: Purpose, Design and Execution*. Cambridge: Cambridge University Press.

Halpin, B. (1975). *Patterns of Animal Disease*. London: Baillière Tindall.

Halse, S.A. & Trevenen, H.J. (1985). Damage to medic pastures by skylarks in north-western Iraq. *Journal of Applied Ecology*, **22**, 337–46.

Harmon, M.E., Bratton, S.P. & White, P.S. (1983). Disturbance and vegetation response in relation to environmental gradients in the Great Smoky Mountains. *Vegetatio*, **55**, 129–39.

Harris, R.B. (1986). Reliability of trend lines obtained from variable counts. *Journal of Wildlife Management*, **50**, 165–71.

Harris, R.B. & Allendorf, F.W. (1989). Genetically effective population size of large mammals: an assessment of estimators. *Conservation Biology*, **3**, 181–91.

Harris, S. (1989). When is a pest a pest? *New Scientist*, **121** (1654), 49–51.

Harris, S. & Rayner, J.M.V. (1986). Models for predicting urban fox (*Vulpes vulpes*) numbers in British cities and their application for rabies control. *Journal of Animal Ecology*, **55**, 593–603.

Harris, S. & Trewhella, W.J. (1988). An analysis of some of the factors affecting dispersal in an urban fox (*Vulpes vulpes*) population. *Journal of Applied Ecology*, **25**, 409–22.

Headley, J.C. (1972a). Defining the economic threshold. In *Pest Control Strategies for the Future*, pp. 100–8. Washington: National Research Council.

Headley, J.C. (1972b). Economics of agricultural pest control. *Annual Review of Entomology*, **17**, 273–86.

Hegdal, P.L. & Blaskiewicz, R.W. (1984). Evaluation of the potential hazard to barn owls of Talon (brodifacoum bait) used to control rats and house mice. *Environmental*

Toxicology and Chemistry, **3**, 167–79.
Hempel, C. G. (1966). *Philosophy of Natural Science*. Englewood Cliffs: Prentice-Hall.
Hengeveld, R. (1989). *Dynamics of Biological Invasions*. London: Chapman & Hall.
Hennemann, W. W. (1983). Relationship among body mass, metabolic rate and the intrinsic rate of natural increase in mammals. *Oecologia*, **56**, 104–8.
Hickson, R. E., Moller, H. & Garrick, A. S. (1986). Poisoning rats on Stewart Island. *New Zealand Journal of Ecology*, **9**, 111–21.
Hill, W. G. (1972). Effective size of populations with overlapping generations. *Theoretical Population Biology*, **3**, 278–89.
Hochberg, M. E. (1989). The potential role of pathogens in biological control. *Nature*, **337**, 262–5.
Holling, C. S. (1959). Some characteristics of simple types of predation and parasitism. *Canadian Entomologist*, **91**, 385–98.
Hone, J. (1977). Dynamics of Loch Ness monsters. *New Scientist*, **76** (1083), 791.
Hone, J. (1980). Probabilities of house mouse (*Mus musculus*) plagues and their use in control. *Australian Wildlife Research*, **7**, 417–20.
Hone, J. (1983). A short-term evaluation of feral pig eradication at Willandra in western New South Wales. *Australian Wildlife Research*, **10**, 269–76.
Hone, J. (1986). Integrative models of poisoning vertebrate pests. In *Proceedings of the Twelfth Vertebrate Pest Conference*, ed. T. P. Salmon, pp. 230–6. Davis: University of California.
Hone, J. (1988a). Evaluation of methods for ground survey of feral pigs and their sign. *Acta Theriologica*, **33**, 451–65.
Hone, J. (1988b). Feral pig rooting in a mountain forest and woodland: distribution, abundance and relationships with environmental variables. *Australian Journal of Ecology*, **13**, 393–400.
Hone, J. (1990). Predator–prey theory and feral pig control, with emphasis on evaluation of shooting from a helicopter. *Australian Wildlife Research*, **17**, 123–30.
Hone, J. (1992a). Modelling of poisoning for vertebrate pest control, with emphasis on poisoning feral pigs. *Ecological Modelling*, **62**, 311–27.
Hone, J. (1992b). Rate of increase and fertility control. *Journal of Applied Ecology*, **29**, 695–8.
Hone, J. & Atkinson, B. (1983). Evluation of fencing to control feral pig movement. *Australian Wildlife Research*, **10**, 499–505.
Hone, J. & Bryant, H. (1981). An examination of feral pig eradication in a hypothetical outbreak of foot and mouth disease at Newcastle, New South Wales. In *Wildlife Diseases of the Pacific Basin and Other Countries. Proceedings of the Fourth International Conference on Wildlife Diseases*, ed. M. E. Fowler, pp. 79–85. Sydney.
Hone, J., Bryant, H., Nicholls, P., Atkinson, W. & Kleba, R. (1985). The acceptance of dyed grain by feral pigs and birds. III. Comparison of intakes of dyed and undyed grain by feral pigs and birds in pig-proof paddocks. *Australian Wildlife Research*, **12**, 447–54.
Hone, J. & Kleba, R. (1984). The toxicity and acceptability of warfarin and 1080 poison to penned feral pigs. *Australian Wildlife Research*, **11**, 103–11.
Hone, J. & Mulligan, H. (1982). *Vertebrate Pesticides*. Science Bulletin 89. Sydney: New South Wales Department of Agriculture.
Hone, J. & Pech, R. (1990). Disease surveillance in wildlife with emphasis on detecting foot and mouth disease in feral pigs. *Journal of Environmental Management*, **31**, 173–84.
Hone, J., Pech, R. & Yip, P. (1992). Estimation of the dynamics and rate of transmission of classical swine fever (hog cholera) in wild pigs. *Epidemiology and Infection*, **108**, 377–86.
Hone, J. & Pedersen, H. (1980). Changes in a feral pig population after poisoning. In

Proceedings of the Ninth Vertebrate Pest Conference, ed. J. P. Clark, pp. 176–82. Davis: University of California.
Hone, J. & Stone, C. P. (1989). A comparison and evaluation of feral pig management in two national parks. *Wildlife Society Bulletin*, **17**, 419–25.
Horn, S. W. (1983). An evaluation of predatory suppression in coyotes using lithium chloride-induced illness. *Journal of Wildlife Management*, **47**, 999–1009.
Hothem, R. L., DeHaven, R. W. & Fairaizl, S. D. (1988). *Bird damage to sunflower in North Dakota, South Dakota, and Minnesota, 1979–1981*. Fish and Wildlife Technical Report 15. Washington: Fish and Wildlife Service.
Howe, T. D., Singer, F. J. & Ackerman, B. B. (1981). Forage relationships of European wild boar invading northern hardwood forest. *Journal of Wildlife Management*, **45**, 748–54.
Hughes, G. & McKinlay, R. G. (1988). Spatial heterogeneity in yield–pest relationships for crop loss assessment. *Ecological Modelling*, **41**, 67–73.
Hughes, R. N. (1979). Optimal diets under the energy maximisation premise: the effects of recognition time and learning. *American Naturalist*, **113**, 209–21.
Hurlbert, S. H. (1984). Pseudoreplication and the design of ecological field experiments. *Ecological Monographs*, **54**, 187–211.
Huson, L. W. (1980). Statistical analysis of comparative field trials of acute rodenticides. *Journal of Hygiene*, **84**, 341–6.
Huston, M. (1979). A general hypothesis of species diversity. *American Naturalist*, **113**, 81–101.
Hygnstrom, S. E. & Craven, S. R. (1988). Electric fences and commercial repellents for reducing deer damage in cornfields. *Wildlife Society Bulletin*, **16**, 291–6.
Inglis, I. R., Thearle, R. J. P. & Isaacson, A. J. (1989). Woodpigeon (*Columba palumbus*) damage to oilseed rape. *Crop Protection*, **8**, 299–309.
Iwasa, Y., Higashi, M. & Yamamura, N. (1981). Prey distribution as a factor determining the choice of optimal foraging strategy. *American Naturalist*, **117**, 710–23.
Izac, A-M. N. & O'Brien, P. (1991). Conflict, uncertainty and risk in feral pig management: the Australian approach. *Journal of Environmental Management*, **32**, 1–18.
Jaksic, F. M. & Yanez, J. L. (1983). Rabbit and fox introductions in Tierra del Fuego: history and assessment of the attempts at biological control of the rabbit infestation. *Biological Conservation*, **26**, 367–74.
Johnson, D. B., Guthery, F. S. & Koerth, N. E. (1989). Grackle damage to grapefruit in the lower Rio Grande valley. *Wildlife Society Bulletin*, **17**, 46–50.
Jones, P. J. (1989). Quelea population dynamics. In *Quelea Quelea: Africa's Bird Pest*, ed. R. L. Bruggers & C. C. H. Elliott, pp. 198–215. Oxford: Oxford University Press.
Jordan, D. J. & Le Feuvre, A. S. (1989). The extent and cause of perinatal lamb mortality in 3 flocks of Merino sheep. *Australian Veterinary Journal*, **66**, 198–201.
Judenko, E. (1973). *Analytical method for assessing yield losses caused by pests on cereal crops with and without pesticides*. London: Centre for Overseas Pest Research.
Juliano, S. A. & Williams, F. M. (1987). A comparison of methods for estimating the functional response parameters of the random predator equation. *Journal of Animal Ecology*, **56**, 641–53.
Kallen, A., Arcuri, P. & Murray, J. D. (1985). A simple model for the spatial spread and control of rabies. *Journal of Theoretical Biology*, **116**, 377–93.
Karr, J. R. & Freemark, K. E. (1985). Disturbance and vertebrates; an integrative perspective. In *The Ecology of Natural Disturbance and Patch Dynamics*, ed. S. T. A. Pickett & P. S. White, pp. 153–68. New York: Academic Press.
Kessell, S. R., Good, R. B. & Hopkins, A. J. M. (1984). Implementation of two new resource

management information systems in Australia. *Environmental Management*, **8**, 251–70.

Kim, K. C. (1983). How to detect and combat exotic pests. In *Exotic Plant Pests and North American Agriculture*, ed. C. L. Wilson & C. L. Graham, pp. 261–319. New York: Academic Press.

King, D. R. (1989). An assessment of the hazard posed to northern quolls (*Dasyurus hallucatus*) by aerial baiting with 1080 to control dingoes. *Australian Wildlife Research*, **16**, 569–74.

King, D. R., Oliver, A. J. & Mead, R. J. (1978). The adaptation of some Western Australian mammals to food plants containing fluoroacetate. *Australian Journal of Zoology*, **26**, 699–712.

Kinnear, J. E., Onus, M. L. & Bromilow, R. N. (1988). Fox control and rock-wallaby population dynamics. *Australian Wildlife Research*, **15**, 435–50.

Kleba, R., Hone, J. & Robards, G. (1985). The acceptance of dyed grain by feral pigs and birds. II. Penned feral pigs. *Ausralian Wildlife Research*, **12**, 51–5.

Kline, M. (1953). *Mathematics in Western Culture*. Oxford: Oxford University Press.

Knight, J. E., Foster, C. L., Howard, V. W. & Schickedanz, J. G. (1986). A pilot test of ultralight aircraft for control of coyotes. *Wildlife Society Bulletin*, **14**, 174–7.

Knight, T. A. & Robinson, F. N. (1978). A possible method of protecting grape crops by using an acoustical device to interfere with communication calls of silvereyes. *Emu*, **78**, 235–6.

Knowles, C. J. (1986). Population recovery of black-tailed prairie dogs following control with zinc phosphide. *Journal of Range Management*, **39**, 249–51.

Knowlton, F. F. (1972). Preliminary interpretations of coyote population mechanics with some management implications. *Journal of Wildlife Management*, **36**, 369–82.

Koenig, W. D. (1988). On determination of viable population size in birds and mammals. *Wildlife Society Bulletin*, **16**, 230–4.

Krebs, C. J. (1985). *Ecology: The Experimental Analysis of Distribution and Abundance*. 3rd edn. San Francisco: Harper & Row.

Krebs, C. J. (1988). The experimental approach to rodent population dynamics. *Oikos*, **52**, 143–9.

Krebs, C. J. (1989). *Ecological Methodology*. New York: Harper & Row.

Krebs, J. R. & McCleery, R. H. (1984). Optimization in behavioural ecology. In *Behavioural Ecology: An Evolutionary Approach*, ed. J. R. Krebs & N. B. Davies, pp. 91–121. Oxford: Blackwell Scientific Publications.

Kruglikov, B. A., Melnik, R. I. & Nalivaiko, V. G. (1985). The part played by wild artiodactyls in carrying the foot-and-mouth disease virus under natural conditions. *Veterinariya*, **8**, 37–8.

Lacki, M. J. & Lancia, R. A. (1983). Changes in soil properties of forests rooted by wild boar. In *Proceedings of the Annual Conference of the Southeast Association of Fish and Wildlife Agencies*, **37**, 228–36.

Lacombe, D., Matton, P. & Cyr, A. (1987). Effect of ornithol on spermatogenesis in red-winged blackbirds. *Journal of Wildlife Management*, **51**, 596–601.

Lange, R. T. & Graham, C. R. (1983). Rabbits and the failure of regeneration in Australian arid zone *Acacia*. *Australian Journal of Ecology*, **8**, 377–81.

Laughlin, R. (1965). Capacity for increase: a useful population statistic. *Journal of Animal Ecology*, **34**, 77–91.

Lawton, J. H., Beddington, J. R. & Bonser, R. (1974). Switching in invertebrate predators. In *Ecological Stability*, ed. M. B. Usher & M. H. Williamson, pp. 141–58. Great Britain: Chapman & Hall.

Lazarus, A. B. & Rowe, F. P. (1982). Reproduction in an island population of Norway rats,

Rattus norvegicus (Berkenhout), treated with an oestrogenic steroid. *Agro-Ecosystems*, **8**, 59–67.

Leader-Williams, N., Walton, D. W. H. & Prince, P. A. (1989). Introduced reindeer on South Georgia: a management dilemma. *Biological Conservation*, **47**, 1–11.

Lefebvre, L. W., Engeman, R. M., Decker, D. G. & Holler, N. R. (1989). Relationship of roof rat population indices with damage to sugarcane. *Wildlife Society Bulletin*, **17**, 41–5.

Lilienfeld, A. M. & Lilienfeld, D. E. (1980). *Foundations of Epidemiology*. 2nd end. Oxford: Oxford University Press.

Liu, W., Levin, S. A. & Iwasa, Y. (1986). Influence of nonlinear incidence rates upon the behavior of SIRS epidemiological models. *Journal of Mathematical Biology*, **23**, 187–204.

Liu, W., Hethcote, H. W. & Levin, S. A. (1987). Dynamical behavior of epidemiological models with nonlinear incidence rates. *Journal of Mathematical Biology*, **25**, 359–80.

Lokemoen, J. T., Doty, H. A., Sharp, D. E. & Neaville, J. E. (1982). Electric fences to reduce mammalian predation on waterfowl nests. *Wildlife Society Bulletin*, **10**, 318–23.

Long, J. L. (1985). Damage to cultivated fruit by parrots in the south of Western Australia. *Australian Wildlife Research*, **12**, 75–80.

Lundberg, P. (1988). Functional response of a small mammalian herbivore: the disc equation revisited. *Journal of Animal Ecology*, **57**, 999–1006.

Lundberg, P. & Danell, K. (1990). Functional response of browsers: tree exploitation by moose. *Oikos*, **58**, 378–84.

Lynch, G. W. & Nass, R. D. (1981). Sodium monofluoroacetate (1080): relation of its use to predation on livestock in western national forests, 1960–78. *Journal of Range Management*, **34**, 421–3.

McAllister, D. M. (1980). *Evaluation in Environmental Planning*. Cambridge, MA: MIT Press.

McCallum, H. I. & Singleton, G. R. (1989). Models to assess the potential of *Capillaria hepatica* to control population outbreaks of house mice. *Parasitology*, **98**, 425–37.

Macdonald, D. (1980). *Rabies and Wildlife*. Oxford: Oxford University Press.

Macfadyen, A. (1975). Some thoughts on the behaviour of ecologists. *Journal of Animal Ecology*, **44**, 351–63.

McIlroy, J. C. (1981). The sensitivity of Australian animals to 1080 poison. I. Instraspecific variation and factors affecting acute toxicity. *Australian Wildlife Research*, **8**, 369–83.

McIlroy, J. C. (1986). The sensitivity of Australian animals to 1080 poison. IX. Comparisons between the major groups of animals, and the potential danger non-target species face from 1080-poisoning campaigns. *Australian Wildlife Research*, **13**, 39–48.

McIlroy, J. C., Braysher, M. & Saunders, G. R. (1989). Effectiveness of a warfarin-poisoning campaign against feral pigs, *Sus scrofa*, in Namadgi National Park, A.C.T. *Australian Wildlife Research*, **16**, 195–202.

McIlroy, J. C., Cooper, R. J., Gifford, E. J., Green, B. F. & Newgrain, K. W. (1986a). The effect on wild dogs, *Canis f. familiaris*, of 1080-poisoning campaigns in Kosciusko National Park, N.S.W. *Australian Wildlife Research*, **13**, 535–44.

McIlroy, J. C. & Gifford, E. J. (1991). Effects on non-target animal populations of a rabbit trail-baiting campaign with 1080 poison. *Wildlife Research*, **18**, 315–25.

McIlroy, J. C., Gifford, E. J. & Cooper, R. J. (1986b). Effect on non-target animal populations of wild dog trail-baiting campaigns with 1080 poison. *Australian Wildlife Research*, **13**, 447–53.

McIlroy, J. C. & Saillard, R. J. (1989). The effect of hunting with dogs on the numbers and movements of feral pigs, *Sus scrofa*, and the subsequent success of poisoning exercises in Namadgi National Park, A.C.T. *Australian Wildlife Research*, **16**, 353–63.

McKay, H. V., Bishop, J. D., Feare, C. J. & Stevens, M. C. (1993). Feeding by brent geese

can reduce yield of oilseed rape. *Crop Protection*, **12**, 101–5.
MacKenzie, D. M. (1990). How Europe is winning its war against rabies. *New Scientist*, **126** (1718), 12–13.
McKillop, I. G. & Sibly, R. M. (1988). Animal behaviour at electric fences and the implications for management. *Mammal Review*, **18**, 91–103.
Mackin, R. (1970). Dynamics of damage caused by wild boar to different agricultural crops. *Acta Theoriologica*, **27**, 447–58.
Maguire, L. A. (1986). Using decision analysis to manage endangered species populations. *Journal of Environmental Management*, **22**, 345–60.
Maguire, L. A., Clark, T. W., Crete, R., Cada, J., Groves, C., Shaffer, M. L. & Seal, U. S. (1988). Black-footed ferret recovery in Montana: a decision analysis. *Wildlife Society Bulletin*, **16**, 111–20.
Maindonald, J. H. (1992). Statistical design, analysis and presentation issues. *New Zealand Journal of Agricultural Research*, **35**, 121–41.
Major, P. F., Dill, L. M. & Eaves, D. M. (1986). Three-dimensional predator–prey interactions: a computer simulation of bird flocks and aircraft. *Canadian Journal of Zoology*, **64**, 2624–33.
May, R. M. (1976). Estimating *r*: a pedagogical note. *American Naturalist*, **110**, 496–99.
May, R. M. (1981a). Models for single populations. In *Theoretical Ecology: Principles and Applications*, 2nd edn, ed. R. M. May, pp. 5–29. Oxford: Blackwell Scientific Publications.
May, R. M. (1981b). Models for two interacting populations. In *Theoretical Ecology: Principles and Applications*, 2nd edn, ed. R. M. May, pp. 78–104. Oxford: Blackwell Scientific Publications.
May, R. M. (1986). When two and two do not make four: nonlinear phenomena in ecology. *Proceedings of the Royal Society London B*, **228**, 241–66.
May, R. M. & Anderson, R. M. (1979). Population biology of infectious diseases: Part II. *Nature*, **280**, 455–61.
May, R. M. & Anderson, R. M. (1983). Epidemiology and genetics in the coevolution of parasites and hosts. *Proceedings of the Royal Society London B*, **219**, 281–313.
May, R. M. & Hassell, M. P. (1988). Population dynamics and biological control. *Philosophical Transactions of the Royal Society London B*, **318**, 129–69.
Maynard Smith, J. (1974). *Models in Ecology*. Cambridge: Cambridge University Press.
Mead, R. J., Oliver, A. J., King, D. R. & Hubach, P. H. (1985). The coevolutionary role of fluoroacetate in plant–animal interactions in Australia. *Oikos*, **44**, 55–60.
Meinzingen, W. W., Bashir, E. S. A., Parker, J. D., Heckel, J. & Elliott, C. C. H. (1989). Lethal control of quelea. In *Quelea Quelea: Africa's Bird Pest*, ed. R. L. Bruggers & C. C. H. Elliott, pp. 293–316. Oxford: Oxford University Press.
Messier, F. & Crete, M. (1985). Moose–wolf dynamics and the natural regulation of moose populations. *Oecologia*, **65**, 503–12.
Millar, J. S. & Zammuto, R. M. (1983). Life histories of mammals: an analysis of life tables. *Ecology*, **64**, 631–5.
Milsom, T. P., Holditch, R. S. & Rochard, J. B. A. (1985). Diurnal use of an airfield and adjacent agricultural habitats by lapwings *Vanellus vanellus*. *Journal of Applied Ecology*, **22**, 313–26.
Mollison, D. (1984). Simplifying simple epidemic models. *Nature*, **310**, 224–5.
Mollison, D. (1986). Modelling biological invasions: chance, explanation, prediction. *Philosophical Transactions of the Royal Society London B*, **314**, 675–93.
Mollison, D. (1991). Dependence of epidemic and population velocities on basic parameters. *Mathematical Biosciences*, **107**, 255–87.

Mollison, D. & Kuulasmaa, K. (1985). Spatial epidemic models: theory and simulations. In *Population Dynamics of Rabies in Wildlife*, ed. P. J. Bacon, pp. 291–309. London: Academic Press.

Montague, C. L., Lefebvre, L. W., Decker, D. G. & Holler, N. R. (1990). Simulation of cotton rat population dynamics and response to rodenticide applications in Florida sugarcane. *Ecological Modelling*, **50**, 177–203.

Morgan, D. R. (1982). Field acceptance of non-toxic and toxic baits by populations of the brushtail possum (*Trichosurus vulpecula* Kerr). *New Zealand Journal of Ecology*, **5**, 36–43.

Morgan, D. R. (1990). Behavioural response of brushtail possums, *Trichosurus vulpecula*, to baits used in pest control. *Australian Wildlife Research*, **17**, 601–13.

Moroney, M. J. (1965). *Facts from Figures*. London: Pelican.

Murdoch, W. W., Chesson, J. & Chesson, P. L. (1985). Biological control in theory and practice. *American Naturalist*, **125**, 344–66.

Murdoch, W. W. & Oaten, A. (1975). Predation and population stability. *Advances in Ecological Research*, **9**, 1–131.

Murray, J. D. & Seward, W. L. (1992). On the spatial spread of rabies among foxes with immunity. *Journal of Theoretical Biology*, **156**, 327–48.

Murray, J. D., Stanley, E. A. & Brown, D. L. (1986). On the spatial spread of rabies among foxes. *Proceedings of the Royal Society London B*, **229**, 111–50.

Murton, R. K. (1968). Some predator–prey relationships in bird damage and population control. In *The Problems of Birds as Pests*, ed. R. K. Murton & E. N. Wright, pp. 157–69. London: Academic Press.

Murton, R. K. & Jones, B. E. (1973). The ecology and economics of damage to Brassicae by wood-pigeons *Columba palumbus*. *Annals of Applied Biology*, **75**, 107–22.

Murton, R. K., Thearle, R. J. P. & Thompson, J. (1972). Ecological studies of the feral pigeon *Columba livia* var. I. Population, breeding biology and methods of control. *Journal of Applied Ecology*, **9**, 835–74.

Murton, R. K., Westwood, N. J. & Isaacson, A. J. (1974). A study of wood-pigeon shooting: the exploitation of a natural animal population. *Journal of Applied Ecology*, **11**, 61–81.

Murton, R. K. & Westwood, N. J. (1976). Birds as pests. In *Applied Biology*, Vol. 1, ed. T. H. Coaker, pp. 89–181. London: Academic Press.

Murua, R. & Rodriguez, J. (1989). An integrated control system for rodents in pine plantations in central Chile. *Journal of Applied Ecology*, **26**, 81–8.

Mutze, G. J. (1989). Effectiveness of strychnine bait trails for poisoning mice in cereal crops. *Australian Wildlife Research*, **16**, 459–65.

Mutze, G. J., Veitch, L. G. & Miller, R. B. (1990). Mouse plagues in South Australian cereal-growing areas. II. An empirical model for prediction of plagues. *Australian Wildlife Research*, **17**, 313–24.

Myers, K. (1954). Studies in the epidemiology of infectious myxomatosis of rabbits. II. Field experiments, August–November 1950, and the first epizootic of myxomatosis in the Riverine plain of south-eastern Australia. *Journal of Hygiene*, **52**, 47–59.

Myers, K. (1970). The rabbit in Australia. In *Proceedings of the Advanced Study Institute on Dynamics of Numbers in Populations*, ed. P. J. den Boer & G. R. Gradwell, pp. 478–506. Wageningen: Centre for Agricultural Publishing and Documentation.

Myers, K. (1986). Introduced vertebrates in Australia, with emphasis on the mammals. In *Ecology of Biological Invasions: An Australian Perspective*, ed. R. H. Groves & J. J. Burdon, pp. 120–36. Canberra: Australian Academy of Science.

Myers, K., Marshall, I. D. & Fenner, F. (1954). Studies in the epidemiology of infectious myxomatosis of rabbits. III. Observations on two succeeding epizootics in Australian wild rabbits on the Riverine Plain of south-eastern Australia 1951–1953. *Journal of*

Hygiene, **52**, 337–62.
Myers, K. & Parker, B.S. (1975). A study of the biology of the wild rabbit in climatically different regions in eastern Australia. VI. Changes in numbers and distribution related to climate and land systems in semiarid north-western New South Wales. *Australian Wildlife Research*, **2**, 11–32.
Myers, K. & Poole, W.E. (1963). A study of the biology of the wild rabbit, *Oryctolagus cuniculus* (L.), in confined populations. IV. The effects of rabbit grazing on sown pastures. *Journal of Ecology*, **51**, 435–51.
Naheed, G. & Khan, J.A. (1989). 'Poison-shyness' and 'bait-shyness' developed by wild rats (*Rattus rattus* L.). I. Methods for eliminating 'shyness' caused by barium carbonate poisoning. *Applied Animal Behaviour Science*, **24**, 89–99.
Nettles, V.F., Corn, J.L., Erickson, G.A. & Jessup, D.A. (1989). A survey of wild swine in the United States for evidence of hog cholera. *Journal of Wildlife Diseases*, **25**, 61–5.
Newsome, A.E., Corbett, L.K., Catling, P.C. & Burt, R.J. (1983). The feeding ecology of the dingo. I. Stomach contents from trapping in south-eastern Australia, and the non-target wildlife also caught in dingo traps. *Australian Wildlife Research*, **10**, 477–86.
Newsome, A.E. & Noble, I.R. (1986). Ecological and physiological characters of invading species. In *Ecology of Biological Invasions: An Australian Perspective*, ed. R.H. Groves & J.J. Burdon, pp. 1–20. Canberra: Australian Academy of Science.
Newsome, A.E., Parer, I. & Catling, P.C. (1989). Prolonged prey suppression by carnivores: predator-removal experiments. *Oecologia*, **78**, 458–67.
Newton, I. (1979). *Population Ecology of Raptors*. Berkhamsted: T. & A.D. Poyser.
Nicholson, A.J. (1933). The balance of animal populations. *Journal of Animal Ecology*, **2**, 132–78.
Norton, G.A. (1976). Analysis of decision making in crop protection. *Agro-Ecosystems*, **3**, 27–44.
Nugent, G. (1990). Forage availability and the diet of fallow deer (*Dama dama*) in the Blue Mountains, Otago, *New Zealand Journal of Ecology*, **13**, 83–95.
O'Brien, P.H. (1985). The impact of feral pigs on livestock production and recent developments in control. *Proceedings of the Australian Society of Animal Production*, **16**, 78–82.
O'Brien, P.H. (1988). The toxicity of sodium monofluoroacetate (compound 1080) to captive feral pigs, *Sus scrofa*. *Australian Wildlife Research*, **15**, 163–70.
O'Brien, P.H., Kleba, R.E., Beck, J.A. & Baker, P.J. (1986). Vomiting by feral pigs after 1080 intoxication: non-target hazard and influence of anti-emetics. *Wildlife Society Bulletin*, **14**, 425–32.
O'Brien, P.H., Lukins, B.S. & Beck, J.A. (1988). Bait type influences the toxicity of sodium monofluoroacetate (compound 1080) to feral pigs. *Australian Wildlife Research*, **15**, 451–7.
O'Gara, B.W., Brawley, K.C., Munoz, J.R. & Henne, D.R. (1983). Predation on domestic sheep on a western Montana ranch. *Wildlife Society Bulletin*, **11**, 253–64.
Okubo, A., Maini, P.K., Williamson, M.H. & Murray, J.D. (1989). On the spatial spread of the grey squirrel in Britain. *Proceedings of the Royal Society London B*, **238**, 113–25.
Oliver, A.J. & King, D.R. (1983). The influence of ambient temperatures on the susceptibility of mice, guinea-pigs and possums to compound 1080. *Australian Wildlife Research*, **10**, 297–301.
Oliver, A.J., Wheeler, S.H. & Gooding, C.D. (1982). Field evaluation of 1080 and pindone oat bait, and the possible decline in effectiveness of poison baiting for the control of the rabbit, *Oryctolagus cuniculus*. *Australian Wildlife Research*, **9**, 125–34.

Orions, G. H. & Pearson, N. E. (1979). On the theory of central place foraging. In *Analysis of Ecological Systems*, ed. D. J. Horn, R. D. Mitchell & G. R. Stairs, pp. 155–77. Columbus: Ohio State University Press.

Otis, D. L. (1989). Damage assessments: estimation methods and sampling design. In *Quelea Quelea: Africa's Bird Pest*, ed. R. L. Bruggers & C. C. H. Elliott, pp. 78–101. Oxford: Oxford University Press.

Otis, D. L., Burnham, K. P., White, G. C. & Anderson, D. R. (1978). Statistical inference from capture data on closed animal populations. *Wildlife Monograph 62.*

Otis, D. L. & Kilburn, C. M. (1988). *Influence of Environmental Factors on Blackbird Damage to Sunflower.* Fish and Wildlife Technical Report 16. Washington: Fish and Wildlife Service.

Parker, B. S., Myers, K. & Caskey, R. L. (1976). An attempt at rabbit control by warren ripping in semi-arid western New South Wales. *Journal of Applied Ecology*, **13**, 353–67.

Parkes, J. P. (1990). Feral goat control in New Zealand. *Biological Conservation*, **54**, 335–48.

Parkes, J. P. & Tustin, K. G. (1985). A reappraisal of the distribution and dispersal of female Himalayan thar in New Zealand. *New Zealand Journal of Ecology*, **8**, 5–10.

Pastoret, P., Brochier, B., Languet, B., Thomas, I., Paquot, A., Bauduin, B., Kieny, M. P., Lecocq, J. P., De Bruyn, J., Costy, F., Antoine, H. & Desmettre, P. (1988). First field trial of fox vaccination against rabies using a vaccinia-rabies recombinant virus. *Veterinary Record*, **123**, 481–3.

Pavlov, P. M. & Hone, J. (1982). The behaviour of feral pigs, *Sus scrofa*, in flocks of lambing ewes. *Australian Wildlife Research*, **9**, 101–9.

Pavlov, P. M., Hone, J., Kilgour, R. J. & Pedersen, H. (1981). Predation by feral pigs on Merino lambs at Nyngan, New South Wales. *Australian Journal of Experimental Agriculture and Animal Husbandry*, **21**, 570–4.

Pearson, E. W. & Caroline, M. (1981). Predator control in relation to livestock losses in central Texas. *Journal of Range Management*, **34**, 435–41.

Pech, R. P. & Hone, J. (1988). A model of the dynamics and control of an outbreak of foot and mouth disease in feral pigs in Australia. *Journal of Applied Ecology*, **25**, 63–77.

Pech, R. P. & McIlroy, J. C. (1990). A model of the velocity of advance of foot and mouth disease in feral pigs. *Journal of Applied Ecology*, **27**, 635–50.

Pech, R. P., Sinclair, A. R. E., Newsome, A. E. & Catling, P. C. (1992). Limits to predator regulation of rabbits in Australia: evidence from predator-removal experiments. *Oecologia*, **89**, 102–12.

Pimm, S. L. (1984). The complexity and stability of ecosystems. *Nature*, **307**, 321–6.

Pimm, S. L. (1987). Determining the effects of introduced species. *Trends in Ecology and Evolution*, **2**, 106–8.

Plant, J. W., Marchant, R., Mitchell, T. D. & Giles, J. R. (1978). Neonatal lamb losses due to feral pig predation. *Australian Veterinary Journal*, **54**, 426–9.

Poche, R. M., Mian, Y., Haque, E. & Sultana, P. (1982). Rodent damage and burrowing characteristics in Bangladesh wheat fields. *Journal of Wildlife Management*, **46**, 139–47.

Pollock, K. H., Nichols, J. D., Brownie, C. & Hines, J. E. (1990). Statistical inference for capture–recapture experiments. *Wildlife Monograph 107.*

Poole, W. E. (1963). Field enclosure experiments on the technique of poisoning the rabbit, *Oryctolagus cuniculus* (L.). III. A study of territorial behaviour and furrow poisoning. *CSIRO Wildlife Research*, **8**, 36–51.

Porter, W. F. (1983). A baited electric fence for controlling deer damage to orchard seedlings. *Wildlife Society Bulletin*, **11**, 325–7.

Prakash, I. (1988). *Rodent Pest Management.* Boca Raton: CRC Press.

Preece, D. A. (1990). R. A. Fisher and experimental design: a review. *Biometrics*, **46**, 925–35.

Pritchard, D. G., Stuart, F. A., Wilesmith, J. W., Cheeseman, C. L., Brewer, J. I., Bode, R. & Sayers, P. E. (1986). Tuberculosis in East Sussex. III. Comparison of post-mortem and clinical methods for the diagnosis of tuberculosis in badgers. *Journal of Hygiene*, **97**, 27–36.

Pybus, M. J. (1988). Rabies and rabies control in striped skunks (*Mephitis mephitis*) in three prairie regions of western north America. *Journal of Wildlife Diseases*, **24**, 434–49.

Quy, R. J., Shepherd, D. S. & Inglis, I. R. (1992). Bait avoidance and effectiveness of anticoagulant rodenticides against warfarin- and difenacoum-resistant populations of Norway rats (*Rattus norvegicus*). *Crop Protection*, **11**, 14–20.

Ralls, K., Brubber, K. & Ballou, J. (1979). Inbreeding and juvenile mortality in small populations of ungulates. *Science*, **206**, 1101–3.

Ralph, C. J. & Maxwell, B. D. (1984). Relative effects of human and feral hog disturbance on a wet forest in Hawaii. *Biological Conservation*, **30**, 291–303.

Real, L. A. (1979). Functional determinants of functional response. *Ecology*, **60**, 481–5.

Redhead, T. D. (1988). Prevention of plagues of house mice in rural Australia. In *Rodent Pest Management*, ed. I. Prakash, pp. 191–205. Boca Raton: CRC Press.

Reed, J. M., Doerr, P. D. & Walters, J. R. (1986). Determining minimum population sizes for birds and mammals. *Wildlife Society Bulletin*, **14**, 255–61.

Rennison, B. D. (1977). Methods for testing rodenticides in the field against rats. *Pesticide Science*, **8**, 405–13.

Rennison, B. D. & Buckle, A. P. (1988). Methods for estimating the losses caused in rice and other crops by rodents. In *Rodent Pest Management*, ed. I. Prakash, pp. 237–59. Boca Raton: CRC Press.

Richards, C. G. J. (1981). Field trials of bromadiolone against infestations of warfarin-resistant *Rattus norvegicus*. *Journal of Hygiene*, **86**, 363–7.

Richards, C. G. J. (1988). Large-scale evaluation of rodent control technologies. In *Rodent Pest Management*, ed. I. Prakash, pp. 269–84. Boca Raton: CRC Press.

Richards, C. G. J. & Huson, L. W. (1985). Towards the optimal use of anticoagulant rodenticides. *Acta Zoologica Fennica*, **173**, 155–7.

Ridpath, M. G. & Waithman, J. (1988). Controlling feral Asian water buffalo in Australia. *Wildlife Society Bulletin*, **16**, 385–90.

Robel, R. J., Dayton, A. D., Henderson, F. R., Meduna, R. L. & Spaeth, C. W. (1981). Relationships between husbandry methods and sheep losses to canine predators. *Journal of Wildlife Management*, **45**, 894–911.

Robertson, G. (1987). Plant dynamics. In *Kangaroos: Their Ecology and Management in the Sheep Rangelands of Australia*, ed. G. Caughley, N. Shepherd & J. Short, pp. 50–68. Cambridge: Cambridge University Press.

Robinson, J. G. & Redford, K. H. (1986). Intrinsic rate of natural increase in Neotropical forest mammals: relationship to phylogeny and diet. *Oecologia*, **68**, 516–20.

Robinson, M. H. & Wheeler, S. H. (1983). A radiotracking study of four poisoning techniques for control of the European rabbit, *Oryctolagus cuniculus* (L.). *Australian Wildlife Research*, **10**, 513–20.

Rogers, D. (1972). Random search and insect population models. *Journal of Animal Ecology*, **41**, 369–83.

Rogers, D. J. (1988). The dynamics of vector-transmitted diseases in human communities. *Philosophical Transactions of the Royal Society London B*, **321**, 513–39.

Ross, J. & Tittensor, A. M. (1986). The establishment and spread of myxomatosis and its effect on rabbit populations. *Philosophical Transactions of the Royal Society London B*, **314**, 599–606.

Ross, J., Tittensor, A. M., Fox, A. P. & Sanders, M. F. (1989). Myxomatosis in farmland rabbit populations in England and Wales. *Epidemiology and Infection*, **103**, 333–57.

Rowe, F. P., Bradfield, A. & Swinney, T. (1985). Pen and field trials of flupropadine against the house mouse (*Mus musculus* L.). *Journal of Hygiene*, **95**, 513–18.

Rowe, F. P., Plant, C. J. & Bradfield, A. (1981). Trials of the anticoagulant rodenticides bromadiolone and difenacoum against the house mouse (*Mus musculus* L.). *Journal of Hygiene*, **87**, 171–7.

Rowley, I. (1957). Field enclosure experiments on the technique of poisoning the rabbit, *Oryctolagus cuniculus* (L.). I. A study of total daily take of bait during free-feeding. *CSIRO Wildlife Research*, **2**, 5–18.

Rowley, I. (1958). Behaviour of a natural rabbit population poisoned with '1080'. *CSIRO Wildlife Research*, **3**, 32–9.

Rowley, I. (1963). Field enclosure experiments on the technique of poisoning the rabbit, *Oryctolagus cuniculus* (L.). IV. A study of shyness in wild rabbits subjected to '1080' poisoning. *CSIRO Wildlife Research*, **8**, 142–53.

Rowley, I. (1968). Studies on the resurgence of rabbit populations after poisoning. *CSIRO Wildlife Research*, **13**, 59–69.

Rowley, I. (1970). Lamb predation in Australia: incidence, predisposing conditions, and the identification of wounds. *CSIRO Wildlife Research*, **15**, 79–123.

Ruesink, W. G. & Kogan, M. (1982). The quantitative basis of pest management: sampling and measurement. In *Introduction to Insect Pest Management*, 2nd edn, ed. R. L. Metcalf & W. H. Luckmann, pp. 315–52. New York: John Wiley & Sons.

Ryan, G. E. & Jones, E. L. (1972). *A Report on the Mouse Plague in the Murrumbidgee and Coleambally Irrigation Areas, 1970*. Sydney: New South Wales Department of Agriculture.

Ryman, N., Bacceus, R., Reuterwall, C. & Smith, M. H. (1981). Effective population size, generation interval, and potential loss of genetic variability in game species under different hunting regimes. *Oikos*, **36**, 257–66.

Saunders, G. (1988). The ecology and management of feral pigs in New South Wales. M.Sc. thesis. Sydney: Macquarie University.

Saunders, G. & Bryant, H. (1988). The evaluation of a feral pig eradication program during a simulated exotic disease outbreak. *Australian Wildlife Research*, **15**, 73–81.

Saunders, G. R. & Cooper, K. (1982). Pesticide contamination of birds in association with mouse plague control. *Emu*, **82**, 227–9.

Saunders, G. R. & Giles, J. R. (1977). A relationship between plagues of the house mouse, *Mus musculus* (Rodentia: Muridae) and prolonged periods of dry weather in south-eastern Australia. *Australian Wildlife Research*, **4**, 241–7.

Saunders, G., Harris, S. & Eason, C. T. (1993). Iophenoxic acid as a quantitative bait marker in foxes. *Wildlife Research*, **20**, 297–302.

Saunders, G., Kay, B. & Parker, B. (1990). Evaluation of a warfarin poisoning programme for feral pigs (*Sus scrofa*). *Australian Wildlife Research*, **17**, 525–33.

Saunders, G. R. & Robards, G. E. (1983). Economic considerations of mouse-plague control in irrigated sunflower crops. *Crop Protection*, **2**, 153–8.

Saunders, I. W. (1980). A model for myxomatosis. *Mathematical Biosciences*, **48**, 1–15.

Schaefer, J. M., Andrews, R. D. & Dinsmore, J. J. (1981). An assessment of coyote and dog predation of sheep in southern Iowa. *Journal of Wildlife Management*, **45**, 883–93.

Schofield, E. K. (1989). Effects of introduced plants and animals on island vegetation: examples from the Galapagos archipelago. *Conservation Biology*, **3**, 227–38.

Scott, M. E. (1988). The impact of infection and disease on animal populations: implications for conservation biology. *Conservation Biology*, **2**, 40–56.

Seaman, J. T., Boulton, J. G. & Carrigan, M. J. (1986). Encephalomyocarditis virus disease of pigs associated with a plague of rodents. *Australian Veterinary Journal*, **63**, 292–4.

Seber, G. A. F. (1982). *The Estimation of Animal Abundance*. 2nd edn. London: Charles Griffin.

Seber, G. A. F. (1986). A review of estimating animal abundance. *Biometrics*, **42**, 267–92.

Shafer, C. L. (1990). *Nature Reserves: Island Theory and Conservation Practice*. Washington: Smithsonian Institution Press.

Shaffer, M. L. (1981). Minimum population sizes for species conservation. *BioScience*, **31**, 131–4.

Shaffer, M. L. (1987). Minimum viable populations: coping with uncertainty. In *Viable Populations for Conservation*, ed. M. E. Soulé, pp. 69–86. Cambridge: Cambridge University Press.

Shaughnessy, P. D. (1980). Influence of Cape fur seals on jackass penguin numbers at Sinclair Island. *South African Journal of Wildlife Research*, **10**, 18–21.

Sheail, J. (1991). The management of an animal population: changing attitudes towards the wild rabbit in Britain. *Journal of Environmental Management*, **33**, 189–203.

Short, J. (1985). The functional response of kangaroos, sheep and rabbits in an arid grazing system. *Journal of Applied Ecology*, **22**, 435–47.

Shorten, M. (1954). The reaction of the brown rat towards changes in its environment. In *Control of Rats and Mice*, Vol. 2, *Rats*, ed. D. Chitty, pp. 307–34. Oxford: Clarendon Press.

Shrum, R. D. & Schein, R. D. (1983). Prediction capabilities for potential epidemics. In *Exotic Plant Pests and North American Agriculture*, ed. C. L. Wilson & C. L. Graham, pp. 419–48. New York: Academic Press.

Shumake, S. A., Kolz, A. L., Reidinger, R. F. & Fall, M. W. (1979). Evaluation of nonlethal electrical barriers for crop protection against rodent damage. In *Vertebrate Pest Control and Management Materials*, ASTM STP 680, ed. J. R. Beck, pp. 29–38. Philadelphia: American Society for Testing and Materials.

Sinclair, A. R. E., Dublin, H. & Borner, M. (1985). Population regulation of Serengeti wildebeest: a test of the food hypothesis. *Oecologia*, **65**, 266–8.

Sinclair, R. G. & Bird, P. L. (1984). The reaction of *Sminthopsis crassicaudata* to meat baits containing 1080: implications for assessing risk to non-target species. *Australian Wildlife Research*, **11**, 501–7.

Sinden, J. A. (1980). Pangloss, pandora and pareto for the aspiring benefit–cost analyst. *Review of Marketing and Agricultural Economics*, **48**, 99–109.

Singer, F. J. (1981). Wild pig populations in the national parks. *Environmental Management*, **5**, 263–70.

Singer, F. J., Swank, W. T. & Clebsch, E. E. C. (1984). Effects of wild pig rooting in a deciduous forest. *Journal of Wildlife Management*, **48**, 464–73.

Singleton, G. R. (1989). Population dynamics of an outbreak of house mice (*Mus domesticus*) in the mallee wheatlands of Australia: hypothesis of plague formation. *Journal of Zoology London*, **219**, 495–515.

Singleton, G. R. & Spratt, D. M. (1986). The effects of *Capillaria hepatica* (Nematoda) on natality and survival to weaning in BALB/c mice. *Australian Journal of Zoology*, **34**, 677–81.

Singleton, G. R., Spratt, D. M., Barker, S. C. & Hodgson, P. F. (1991b). The geographical distribution and host range of *Capillaria hepatica* (Bancroft) (Nematoda) in Australia. *International Journal of Parasitology*, **21**, 945–57.

Singleton, G. R., Twigg, L. E., Weaver, K. E. & Kay, B. J. (1991a). Evaluation of bromadiolone against house mouse (*Mus domesticus*) populations in irrigated soybean

crops. II. Economics. *Wildlife Research*, **18**, 275–83.
Skalski, J. R. & Robson, D. S. (1992). *Techniques for Wildlife Investigations: Design and Analysis of Capture Data*. London: Academic Press.
Skalski, J. R., Robson, D. S. & Matsuzaki, C. L. (1983). Competing probabilistic models for catch–effort relationships in wildlife censuses. *Ecological Modelling*, **19**, 299–307.
Skeat, A. (1990). Feral buffalo in Kakadu National Park: survey methods, population dynamics and control. M.Appl.Sci. thesis. Canberra: University of Canberra.
Smart, C. W. & Giles, R. H. (1973). A computer model of wildlife rabies epizootics and an analysis of incidence patterns. *Wildlife Diseases*, **61**, 1–89.
Smith, A. D. M. (1985). A continuous time deterministic model of temporal rabies. In *Population Dynamics of Rabies in Wildlife*, ed. P. J. Bacon, pp. 131–46. London: Academic Press.
Smith, G. C. & Harris, S. (1989). The control of rabies in urban fox populations. In *Mammals as Pests*, ed. R. J. Putman, pp. 209–24. London: Chapman & Hall.
Smith, G. C. & Harris, S. (1991). Rabies in urban foxes (*Vulpes vulpes*) in Britain: the use of a spatial stochastic simulation model to examine the pattern of spread and evaluate the efficacy of different control regimes. *Philosophical Transactions of the Royal Society London B*, **334**, 459–79.
Smith, R. H., Neff, D. J. & Woolsey, N. G. (1986). Pronghorn response to coyote control: a benefit: cost analysis. *Wildlife Society Bulletin*, **14**, 226–31.
Snedecor, G. W. & Cochran, W. G. (1967). *Statistical Methods*, 6th edn. Ames: Iowa State University Press.
Snowdon, W. A. (1968). The susceptibility of some Australian fauna to infection with foot and mouth disease virus. *Australian Journal of Experimental Biology and Medical Science*, **46**, 667–87.
Snyder, R. (1984). Basic concepts of the dose–response relationship. In *Assessment and Management of Chemical Risks*, ed. J. V. Rodricks & R. G. Tardiff, pp. 37–56. Washington: American Chemical Society.
Solman, V. E. F. (1973). Birds and aircraft. *Biological Conservation*, **5**, 79–86.
Solman, V. E. F. (1978). Gulls and aircraft. *Environmental Conservation*, **5**, 277–80.
Soulé, M. E. (1987a). *Viable Populations for Conservation*. Cambridge: Cambridge University Press.
Soulé, M. E. (1987b). Introduction. In *Viable Populations for Conservation*, ed. M. E. Soulé, pp. 1–10. Cambridge: Cambridge University Press.
Soulé, M. E. (1987c). Where do we go from here? In *Viable Populations for Conservation*, ed. M. E. Soulé, pp. 175–83. Cambridge: Cambridge University Press.
Southwood, T. R. E. (1988). Tactics, strategies and templets. *Oikos*, **52**, 3–18.
Southwood, T. R. E. & Norton, G. (1973). Economic aspects of pest management strategies and decisions. In *Insects: Studies in Pest Management*, P. W. Geier, L. R. Clark, D. J. Anderson & H. A. Nix, pp. 168–84. Canberra: Ecological Society of Australia.
Spratt, D. M. (1990). The role of helminths in the biological control of mammals. *International Journal for Parasitology*, **20**, 543–50.
Spurr, E. B. (1979). A theoretical assessment of the ability of bird species to recover from an imposed reduction in numbers, with particular reference to 1080 poisoning. *New Zealand Journal of Ecology*, **2**, 46–63.
Stamps, J. A., Buechner, M. & Krishnan, V. V. (1987). The effects of edge permeability and habitat geometry on emigration from patches of habitat. *American Naturalist*, **129**, 533–52.
Starfield, A. M. & Bleloch, A. L. (1986). *Building Models for Conservation and Wildlife*

Management. New York: Macmillan.
Steel, R. G. & Torrie, J. H. (1960). *Principles and Procedures of Statistics*. New York: McGraw-Hill.
Stenseth, N. C. (1977). On the importance of spatio-temporal heterogeneity for the population dynamics of rodents: towards a theoretical foundation of rodent control. *Oikos*, **29**, 545–52.
Stenseth, N. C. (1981). How to control pest species: application of models from the theory of island biogeography in formulating pest control strategies. *Journal of Applied Ecology*, **18**, 773–94.
Stenseth, N. C. (1984). Why mathematical models in evolutionary ecology? In *Trends in Ecological Research for the 1980s*, ed. J. H. Cooley & F. B. Golley, pp. 239–87. New York: Plenum Press.
Stenseth, N. C. & Hansson, L. (1981). The importance of population dynamics in heterogeneous landscapes: management of vertebrate pests and some other animals. *Agro-Ecosystems*, **7**, 187–211.
Stephens, D. W. & Krebs, J. R. (1986). *Foraging Theory*. Princeton: Princeton University Press.
Sterner, R. T. & Shumake, S. A. (1978). Coyote damage-control research: a review and analysis. In *Coyotes: Biology, Behavior and Management*, ed. M. Bekoff, pp. 297–325. New York: Academic Press.
Stickley, A. R., Otis, D. L. & Palmer, D. T. (1979). Evaluation and results of a survey of blackbird and mammal damage to mature field corn over a large (three-state) area. In *Vertebrate Pest Control and Management Materials*, ASTM STP 680, ed. J. R. Beck, pp. 169–77. Philadelphia: American Society for Testing and Materials.
Stone, C. P. (1985). Alien animals in Hawai'i's native ecosystems: towards controlling the adverse effects of introduced vertebrates. In *Hawai'i's Terrestrial Ecosystems: Preservation and Management*, ed. C. P. Stone & J. M. Scott, pp. 251–97. Hawaii: Cooperative National Park Resources Studies Unit, University of Hawaii.
Straub, R. W. (1989). Red-winged blackbird damage to sweet corn in relation to infestations of European corn borer (Lepidoptera: Pyralidae). *Journal of Economic Entomology*, **82**, 1406–10.
Stuart, F. A. & Wilesmith, J. W. (1988). Tuberculosis in badgers: a review. *Reviews of Science and Technology Office International Epizootic*, **7**, 929–35.
Sudman, S., Sirken, M. G. & Cowan, C. D. (1988). Sampling rare and elusive populations. *Science*, **240**, 991–6.
Sukumar, R. (1991). The management of large mammals in relation to male strategies and conflict with people. *Biological Conservation*, **55**, 93–102.
Sullivan, T. P., Crump, D. R. & Sullivan, D. S. (1988a). Use of predator odors as repellents to reduce feeding damage by herbivores. III. Montane and meadow voles (*Microtus montanus* and *Microtus pennsylvanicus*). *Journal of Chemical Ecology*, **14**, 363–77.
Sullivan, T. P., Crump, D. R. & Sullivan, D. S. (1988b). Use of predator odors as repellents to reduce feeding damage by herbivores. IV. Northern pocket gophers (*Thomomys talpoides*). *Journal of Chemical Ecology*, **14**, 379–89.
Summers, R. W. (1990). The effect on winter wheat of grazing by brent geese *Branta bernicla*. *Journal of Applied Ecology*, **27**, 821–33.
Sumption, K. J. & Flowerdew, J. R. (1985). The ecological effects of the decline in rabbits (*Oryctolagus cuniculus* L.) due to myxomatosis. *Mammal Review*, **15**, 151–86.
Taylor, D. & Katahira, L. (1988). Radio telemetry as an aid in eradicating remnant feral goats. *Wildlife Society Bulletin*, **16**, 297–9.
Taylor, R. D. & Martin, R. B. (1987). Effects of veterinary fences on wildlife conservation in

Zimbabwe. *Environmental Management*, **11**, 327–34.
Taylor, R. H. & Thomas, B. W. (1993). Rats eradicated from rugged Breaksea Island (170 ha), Fiordland, New Zealand. *Biological Conservation*, **65**, 191–8.
Taylor, R. J. (1977). The value of clumping to prey. Experiments with a mammalian predator. *Oecologia*, **30**, 285–94.
Taylor, R. J. (1984). *Predation*. New York: Chapman & Hall.
Temme, M. & Jackson, W. B. (1979). Criteria for trap evaluation. In *Vertebrate Pest Control and Management Materials*, ASTM STP 680, ed. J. R. Beck, pp. 58–67. Philadelphia: American Society for Testing and Materials.
Terrill, C. E. (1986). Trends of predator losses of sheep and lambs from 1940 through 1985. In *Proceedings of the Twelfth Vertebrate Pest Conference*, ed. T. P. Salmon, pp. 347–51. Davis: University of California.
Thompson, W. R. (1947). Use of moving averages and interpolation to estimate median-effective doses. I. Fundamental formulas, estimation of error, and relation to other methods. *Bacteriology Reviews*, **11**, 115–45.
Thomson, P. C. (1986). The effectiveness of aerial baiting for the control of dingoes in north-western Australia. *Australian Wildlife Research*, **13**, 165–76.
Thornes, J. B. (1988). Erosional equilibria under grazing. In *Conceptual Issues in Environmental Archaeology*, ed. J. Bintliff, D. A. Davidson, & E. G. Grant, pp. 193–210. Edinburgh; Edinburgh University Press.
Thornes, J. B. (1990). The interaction of erosional and vegetational dynamics in land degradation: spatial outcomes. In *Vegetation and Erosion*, ed. J. B. Thornes, pp. 41–53. London: John Wiley & Sons.
Thornton, P. S. (1988). Density and distribution of badgers in south-west England: a predictive model. *Mammal Review*, **18**, 11–23.
Tietjen, H. P. & Matschke, G. H. (1982). Aerial prebaiting for management of prairie dogs with zinc phosphide. *Journal of Wildlife Management*, **46**, 1108–12.
Timm, R. M. (1983). *Prevention and control of wildlife damage*. Lincoln: University of Nebraska.
Tisdell, C. (1982). Wild Pigs: Environmental Pest or Economic Resource? Sydney: Pergamon Press.
Tisdell, C. A. & Auld, B. A. (1988). Evaluation of biological control projects. In *Proceedings of VII International Symposium on Biological Control of Weeds*, ed. E. S. Delfosse, pp. 93–110. Rome.
Tisdell, C. A., Auld, B. A. & Menz, K. M. (1984a). On assessing the value of biological control of weeds. *Protection Ecology*, **6**, 169–79.
Tisdell, C. A., Auld, B. A. & Menz, K. M. (1984b). Crop loss elasticity in relation to weed density and control. *Agricultural Systems*, **13**, 161–6.
Tobin, M. E., Koehler, A. E., Sugihara, R. T., Ueunten, G. R. & Yamaguchi, A. M. (1993). Effects of trapping on rat populations and subsequent damage and yields of macadamia nuts. *Crop Protection*, **12**, 243–8.
Tome, M. W. (1988). Optimal foraging: food patch depletion by ruddy ducks. *Oecologia*, **76**, 27–36.
Tomich, P. Q. (1986). *Mammals in Hawai'i*, 2nd edn. Honolulu: Bishop Museum Press.
Travis, C. C. & Lenhart, S. M. (1987). Eradication of infectious diseases in heterogeneous populations. *Mathematical Biosciences*, **83**, 191–8.
Trewhella, W. J. & Harris, S. (1988). A simulation model of the pattern of dispersal in urban fox (*Vulpes vulpes*) populations and its application for rabies control. *Journal of Applied Ecology*, **25**, 435–50.
Trewhella, W. J., Harris, S. & McAllister, F. E. (1988). Dispersal distance, home-

range size and population density in the red fox (*Vulpes vulpes*): a quantitative analysis. *Journal of Applied Ecology*, **25**, 423–34.

Trewhella, W.J., Harris, S., Smith, G.C. & Nadian, A.K. (1991). A field trial evaluating bait uptake by an urban fox (*Vulpes vulpes*) population. *Journal of Applied Ecology*, **28**, 454–66.

Trout, R.C., Ross, J., Tittensor, A.M. & Fox, A.P. (1992). The effect on a British wild rabbit population (*Oryctolagus cuniculus*) of manipulating myxomatosis. *Journal of Applied Ecology*, **29**, 679–86.

Twigg, L.E. & King, D.R. (1991). The impact of fluoroacetate-bearing vegetation on native Australian fauna: a review. *Oikos*, **61**, 412–30.

Twigg, L.E., Singleton, G.R. & Kay, B.J. (1991). Evaluation of bromadiolone against house mouse (*Mus domesticus*) populations in irrigated soybean crops. I. Efficacy of control. *Wildlife Research*, **18**, 265–74.

Usher, M.B., Kruger, F.J., Macdonald, I.A.W., Loope, L.L. & Brockie, R.E. (1988). The ecology of biological invasions into nature reserves: an introduction. *Biological Conservation*, **44**, 1–8.

van den Bosch, F., Hengeveld, R. & Metz, J.A.J. (1992). Analysing the velocity of animal range expansion. *Journal of Biogeography*, **19**, 135–50.

van den Bosch, F., Metz, J.A.J. & Diekmann, O. (1990). The velocity of spatial population expansion. *Journal of Mathematical Biology*, **28**, 529–65.

van Rensburg, P.J.J., Skinner, J.D. & van Aarde, R.J. (1987). Effects of feline panleucopaenia on the population characteristics of feral cats on Marion Island. *Journal of Applied Ecology*, **24**, 63–73.

van Tets, G.F. (1969). Quantitative and qualitative changes in habitat and avifauna at Sydney airport. *CSIRO Wildlife Research*, **14**, 117–28.

van Tets, G.F., Vestjens, W.J.M., D'Andria, A.H. & Barker, R. (1977). *Guide to the Recognition and Reduction of Aerodrome Bird Hazards*. Canberra: Australian Government Publishing Service.

Vitale, A.F. (1989). Changes in the anti-predator responses of wild rabbits, *Oryctolagus cuniculus* (L.), with age and experience. *Behaviour*, **110**, 47–61.

Waage, J.K. & Greathead, D.J. (1988). Biological control: challenges and opportunities. *Philosophical Transactions of the Royal Society London B*, **318**, 111–28.

Wade, D.A. (1978). Coyote damage: a survey of its nature and scope, control measures and their application. In *Coyotes: Biology, Behavior and Management*, ed. M. Bekoff, pp. 347–68. New York: Academic Press.

Wade, D.A. & Bowns, J.E. (1982). *Procedures for Evaluating Predation on Livestock and Wildlife*. Bulletin No. B-1429 Texas: Texas A & M University.

Wagner, F.H. (1972). *Coyotes and Sheep. Some Thoughts on Ecology, Economics and Ethics*. 44th Honor lecture. Logan: Utah State University.

Wagner, F.H. (1975). The predator-control scene as of 1974. *Journal of Range Management*, **28**, 4–10.

Wagner, F.H. & Pattison, L.G. (1973). Analysis of existing data on sheep loss and predator activity. In *Final Report, Four Corners Regional Commission. Predator Control Study*. Logan: Utah State University.

Wakeley, J.S. & Mitchell, R.C. (1981). Blackbird damage to ripening field corn in Pennsylvania. *Wildlife Society Bulletin*, **9**, 52–5.

Walker, B.H. & Norton, G.A. (1982). Applied ecology: towards a positive approach. II. Applied ecological analysis. *Journal of Environmental Management*, **14**, 325–42.

Walters, C. (1986). *Adaptive Management of Renewable Resources*. New York: Macmillan.

Waters, W.E. (1955). Sequential sampling in forest insect surveys. *Forest Science*, **1**, 68–79.

Weatherhead, P. J., Tinker, S. & Greenwood, H. (1982). Indirect assessment of avian damage to agriculture. *Journal of Applied Ecology*, **19**, 773–82.

Weil, C. S. (1952). Tables for convenient calculation of median-effective doses (LD_{50} or ED_{50}) and instructions in their use. *Biometrics*, **8**, 249–63.

Whysong, G. L. & Brady, W. W. (1987). Frequency sampling and type II errors. *Journal of Range Management*, **40**, 472–4.

Wiens, J. A. & Dyer, M. I. (1975). Simulation modelling of red-winged blackbird impact on grain crops. *Journal of Applied Ecology*, **12**, 63–82.

Williams, J. M., Bell, J., Ross, W. D. & Broad, T. M. (1986). Rabbit (*Oryctolagus cuniculus*) control with a single application of 50 ppm brodifacoum cereal baits. *New Zealand Journal of Ecology*, **9**, 123–36.

Williamson, M. H. & Brown, K. C. (1986). The analysis and modelling of British invasions. *Philosophical Transactions of the Royal Society London B*, **314**, 505–22.

Windberg, L. A. & Knowlton, F. F. (1990). Relative vulnerability of coyotes to some capture procedures. *Wildlife Society Bulletin*, **18**, 282–90.

Wittgenstein, L. (1967). *Philosophical Investigations*, 3rd edn. Oxford: Basil Blackwell.

Wood, B. J. & Liau, S. S. (1984). A long-term study of *Rattus tiomanicus* populations in an oil palm plantation in Johore, Malaysia. *Journal of Applied Ecology*, **21**, 465–72.

Woodall, P. F. (1983). Distribution and population dynamics of dingoes (*Canis familiaris*) and feral pigs (*Sus scrofa*) in Queensland, 1945–1976. *Journal of Applied Ecology*, **20**, 85–95.

Woods, A. (1974). *Pest Control: A Survey*. London: McGraw Hill.

Woollons, R. C. & Whyte, A. G. D. (1988). Multiple covariance: its utility in analysing forest fertilizer experiments. *Forest Ecology and Management*, **25**, 59–72.

Woronecki, P. P. & Dolbeer, R. A. (1980). The influence of insects in bird damage control. In *Proceedings Ninth Vertebrate Pest Conference*, ed. J. P. Clark, pp. 53–9. Davis: University of California.

Woronecki, P. P., Stehn, R. A. & Dolbeer, R. A. (1980). Compensatory response of maturing corn kernels following simulated damage by birds. *Journal of Applied Ecology*, **17**, 737–46.

Wright, E. N. (1968). Modification of the habitat as a means of bird control. In *The Problems of Birds as Pests*, ed. R. K. Murton & E. N. Wright, pp. 97–105. London: Academic Press.

Yip, P. (1989). Estimating the initial relative infection rate for a stochastic epidemic model. *Theoretical Population Biology*, **36**, 202–13.

Yom-Tov, Y. (1980). The timing of pest control operations in relation to secondary poisoning prevention. *Biological Conservation*, **18**, 143–7.

Author index

Subject index